EXPOSURE SCIENCE
in the 21st Century

A VISION AND A STRATEGY

Committee on Human and Environmental
Exposure Science in the 21st Century

Board on Environmental Studies and Toxicology

Division on Earth and Life Studies

NATIONAL RESEARCH COUNCIL
OF THE NATIONAL ACADEMIES

THE NATIONAL ACADEMIES PRESS
Washington, D.C.
www.nap.edu

THE NATIONAL ACADEMIES PRESS   500 Fifth Street, NW   Washington, DC 20001

NOTICE: The project that is the subject of this report was approved by the Governing Board of the National Research Council, whose members are drawn from the councils of the National Academy of Sciences, the National Academy of Engineering, and the Institute of Medicine. The members of the committee responsible for the report were chosen for their special competences and with regard for appropriate balance.

This project was supported by Contract EP-C-09-003 between the National Academy of Sciences and U.S. Environmental Protection Agency. The project was also supported by the National Institute of Environmental Health Sciences through this contract. Any opinions, findings, conclusions, or recommendations expressed in this publication are those of the authors and do not necessarily reflect the view of the organizations or agencies that provided support for this project.

International Standard Book Number-13: 978-0-309-26468-6
International Standard Book Number-10: 0-309-26468-5
Library of Congress Control Number: 2012949228

Additional copies of this report are available for sale from the National Academies Press, 500 Fifth Street, NW, Keck 360, Washington, DC 20001; (800) 624-6242 or (202) 334-3313; http://www.nap.edu/.

Copyright 2012 by the National Academy of Sciences. All rights reserved.

Printed in the United States of America

THE NATIONAL ACADEMIES
Advisers to the Nation on Science, Engineering, and Medicine

The **National Academy of Sciences** is a private, nonprofit, self-perpetuating society of distinguished scholars engaged in scientific and engineering research, dedicated to the furtherance of science and technology and to their use for the general welfare. Upon the authority of the charter granted to it by the Congress in 1863, the Academy has a mandate that requires it to advise the federal government on scientific and technical matters. Dr. Ralph J. Cicerone is president of the National Academy of Sciences.

The **National Academy of Engineering** was established in 1964, under the charter of the National Academy of Sciences, as a parallel organization of outstanding engineers. It is autonomous in its administration and in the selection of its members, sharing with the National Academy of Sciences the responsibility for advising the federal government. The National Academy of Engineering also sponsors engineering programs aimed at meeting national needs, encourages education and research, and recognizes the superior achievements of engineers. Dr. Charles M. Vest is president of the National Academy of Engineering.

The **Institute of Medicine** was established in 1970 by the National Academy of Sciences to secure the services of eminent members of appropriate professions in the examination of policy matters pertaining to the health of the public. The Institute acts under the responsibility given to the National Academy of Sciences by its congressional charter to be an adviser to the federal government and, upon its own initiative, to identify issues of medical care, research, and education. Dr. Harvey V. Fineberg is president of the Institute of Medicine.

The **National Research Council** was organized by the National Academy of Sciences in 1916 to associate the broad community of science and technology with the Academy's purposes of furthering knowledge and advising the federal government. Functioning in accordance with general policies determined by the Academy, the Council has become the principal operating agency of both the National Academy of Sciences and the National Academy of Engineering in providing services to the government, the public, and the scientific and engineering communities. The Council is administered jointly by both Academies and the Institute of Medicine. Dr. Ralph J. Cicerone and Dr. Charles M. Vest are chair and vice chair, respectively, of the National Research Council.

www.national-academies.org

**COMMITTEE ON HUMAN AND ENVIRONMENTAL
EXPOSURE SCIENCE IN THE 21ST CENTURY**

Members

KIRK R. SMITH (*Chair*), University of California, Berkeley, CA
PAUL J. LIOY (*Vice Chair*), University of Medicine and Dentistry of New Jersey, Piscataway, NJ
TINA BAHADORI, American Chemistry Council, Washington, DC (resigned March 2012)
TIMOTHY BUCKLEY, Ohio State University, Columbus, OH (resigned May 2012)
RICHARD T. DI GIULIO, Duke University, Durham, NC
J. PAUL GILMAN, Covanta Energy Corporation, Fairfield, NJ
MICHAEL JERRETT, University of California, Berkeley, CA
DEAN JONES, Emory University, Atlanta, GA (resigned June 2012)
PETROS KOUTRAKIS, Harvard School of Public Health, Boston, MA
THOMAS E. MCKONE, University of California, Berkeley, CA
JAMES T. ORIS, Miami University, Oxford, OH
AMANDA D. RODEWALD, Ohio State University, Columbus, OH
SUSAN L. SANTOS, University of Medicine and Dentistry of New Jersey, Piscataway, NJ
RICHARD SHARP, Cleveland Clinic, Cleveland, OH
GINA SOLOMON, California Environmental Protection Agency, Sacramento, CA
JUSTIN G. TEEGUARDEN, Pacific Northwest National Laboratory, Richland, WA
DUNCAN C. THOMAS, University of Southern California, Los Angeles, CA
THOMAS G. THUNDAT, University of Alberta, Edmonton, AB, Canada
SACOBY M. WILSON, University of Maryland, College Park, MD

Staff

EILEEN N. ABT, Project Director
KEEGAN SAWYER, Program Officer (through September 2011)
KERI SCHAFFER, Research Associate
NORMAN GROSSBLATT, Senior Editor
MIRSADA KARALIC-LONCAREVIC, Manager, Technical Information Center
RADIAH ROSE, Manager, Editorial Projects
ORIN LUKE, Senior Program Assistant (through June 2011)
TAMARA DAWSON, Program Associate

Sponsor

U.S. ENVIRONMENTAL PROTECTION AGENCY
NATIONAL INSTITUTE OF ENVIRONMENTAL HEALTH SCIENCES

BOARD ON ENVIRONMENTAL STUDIES AND TOXICOLOGY

Members

ROGENE F. HENDERSON (*Chair*), Lovelace Respiratory Research Institute, Albuquerque, NM
PRAVEEN AMAR, Clean Air Task Force, Boston, MA
MICHAEL J. BRADLEY, M.J. Bradley & Associates, Concord, MA
JONATHAN Z. CANNON, University of Virginia, Charlottesville
GAIL CHARNLEY, HealthRisk Strategies, Washington, DC
FRANK W. DAVIS, University of California, Santa Barbara
RICHARD A. DENISON, Environmental Defense Fund, Washington, DC
CHARLES T. DRISCOLL, JR., Syracuse University, New York
H. CHRISTOPHER FREY, North Carolina State University, Raleigh
RICHARD M. GOLD, Holland & Knight, LLP, Washington, DC
LYNN R. GOLDMAN, George Washington University, Washington, DC
LINDA E. GREER, Natural Resources Defense Council, Washington, DC
WILLIAM E. HALPERIN, University of Medicine and Dentistry of New Jersey, Newark
PHILIP K. HOPKE, Clarkson University, Potsdam, NY
HOWARD HU, University of Michigan, Ann Arbor
SAMUEL KACEW, University of Ottawa, Ontario
ROGER E. KASPERSON, Clark University, Worcester, MA
THOMAS E. MCKONE, University of California, Berkeley
TERRY L. MEDLEY, E.I. du Pont de Nemours & Company, Wilmington, DE
JANA MILFORD, University of Colorado at Boulder, Boulder
FRANK O'DONNELL, Clean Air Watch, Washington, DC
RICHARD L. POIROT, Vermont Department of Environmental Conservation, Waterbury
KATHRYN G. SESSIONS, Health and Environmental Funders Network, Bethesda, MD
JOYCE S. TSUJI, Exponent Environmental Group, Bellevue, WA

Senior Staff

JAMES J. REISA, Director
DAVID J. POLICANSKY, Scholar
RAYMOND A. WASSEL, Senior Program Officer for Environmental Studies
ELLEN K. MANTUS, Senior Program Officer for Risk Analysis
SUSAN N.J. MARTEL, Senior Program Officer for Toxicology
EILEEN N. ABT, Senior Program Officer
MIRSADA KARALIC-LONCAREVIC, Manager, Technical Information Center
RADIAH ROSE, Manager, Editorial Projects

OTHER REPORTS OF THE
BOARD ON ENVIRONMENTAL STUDIES AND TOXICOLOGY

Science for Environmental Protection: The Road Ahead (2012)
A Research Strategy for Environmental, Health, and Safety Aspects of Engineered Nanomaterials (2012)
Macondo Well–Deepwater Horizon Blowout: Lessons for Improving Offshore Drilling Safety (2012)
Feasibility of Using Mycoherbicides for Controlling Illicit Drug Crops (2011)
Improving Health in the United States: The Role of Health Impact Assessment (2011)
A Risk-Characterization Framework for Decision-Making at the Food and Drug Administration (2011)
Review of the Environmental Protection Agency's Draft IRIS Assessment of Formaldehyde (2011)
Toxicity-Pathway-Based Risk Assessment: Preparing for Paradigm Change (2010)
The Use of Title 42 Authority at the U.S. Environmental Protection Agency (2010)
Review of the Environmental Protection Agency's Draft IRIS Assessment of Tetrachloroethylene (2010)
Hidden Costs of Energy: Unpriced Consequences of Energy Production and Use (2009)
Contaminated Water Supplies at Camp Lejeune—Assessing Potential Health Effects (2009)
Review of the Federal Strategy for Nanotechnology-Related Environmental, Health, and Safety Research (2009)
Science and Decisions: Advancing Risk Assessment (2009)
Phthalates and Cumulative Risk Assessment: The Tasks Ahead (2008)
Estimating Mortality Risk Reduction and Economic Benefits from Controlling Ozone Air Pollution (2008)
Respiratory Diseases Research at NIOSH (2008)
Evaluating Research Efficiency in the U.S. Environmental Protection Agency (2008)
Hydrology, Ecology, and Fishes of the Klamath River Basin (2008)
Applications of Toxicogenomic Technologies to Predictive Toxicology and Risk Assessment (2007)
Models in Environmental Regulatory Decision Making (2007)
Toxicity Testing in the Twenty-first Century: A Vision and a Strategy (2007)
Sediment Dredging at Superfund Megasites: Assessing the Effectiveness (2007)
Environmental Impacts of Wind-Energy Projects (2007)
Scientific Review of the Proposed Risk Assessment Bulletin from the Office of Management and Budget (2007)
Assessing the Human Health Risks of Trichloroethylene: Key Scientific Issues (2006)
New Source Review for Stationary Sources of Air Pollution (2006)
Human Biomonitoring for Environmental Chemicals (2006)
Health Risks from Dioxin and Related Compounds: Evaluation of the EPA Reassessment (2006)

Fluoride in Drinking Water: A Scientific Review of EPA's Standards (2006)
State and Federal Standards for Mobile-Source Emissions (2006)
Superfund and Mining Megasites—Lessons from the Coeur d'Alene River Basin (2005)
Health Implications of Perchlorate Ingestion (2005)
Air Quality Management in the United States (2004)
Endangered and Threatened Species of the Platte River (2004)
Atlantic Salmon in Maine (2004)
Endangered and Threatened Fishes in the Klamath River Basin (2004)
Cumulative Environmental Effects of Alaska North Slope Oil and Gas Development (2003)
Estimating the Public Health Benefits of Proposed Air Pollution Regulations (2002)
Biosolids Applied to Land: Advancing Standards and Practices (2002)
The Airliner Cabin Environment and Health of Passengers and Crew (2002)
Arsenic in Drinking Water: 2001 Update (2001)
Evaluating Vehicle Emissions Inspection and Maintenance Programs (2001)
Compensating for Wetland Losses Under the Clean Water Act (2001)
A Risk-Management Strategy for PCB-Contaminated Sediments (2001)
Acute Exposure Guideline Levels for Selected Airborne Chemicals (twelve volumes, 2000-2012)
Toxicological Effects of Methylmercury (2000)
Strengthening Science at the U.S. Environmental Protection Agency (2000)
Scientific Frontiers in Developmental Toxicology and Risk Assessment (2000)
Ecological Indicators for the Nation (2000)
Waste Incineration and Public Health (2000)
Hormonally Active Agents in the Environment (1999)
Research Priorities for Airborne Particulate Matter (four volumes, 1998-2004)
The National Research Council's Committee on Toxicology: The First 50 Years (1997)
Carcinogens and Anticarcinogens in the Human Diet (1996)
Upstream: Salmon and Society in the Pacific Northwest (1996)
Science and the Endangered Species Act (1995)
Wetlands: Characteristics and Boundaries (1995)
Biologic Markers (five volumes, 1989-1995)
Science and Judgment in Risk Assessment (1994)
Pesticides in the Diets of Infants and Children (1993)
Dolphins and the Tuna Industry (1992)
Science and the National Parks (1992)
Human Exposure Assessment for Airborne Pollutants (1991)
Rethinking the Ozone Problem in Urban and Regional Air Pollution (1991)
Decline of the Sea Turtles (1990)

Copies of these reports may be ordered from the National Academies Press
(800) 624-6242 or (202) 334-3313
www.nap.edu

Preface

Over the last decade, advances in tools and technologies—sensor systems, analytic methods, molecular technologies, computational tools, and bioinformatics—have provided opportunities for improving the collection of exposure-science information leading to the potential for better human health and ecosystem protection. Recognizing the need for a prospective examination of exposure science, the U.S. Environmental Protection Agency and the National Institute of Environmental Health Sciences asked the National Research Council to perform an independent study to develop a long-range vision and a strategy for implementing the vision over the next 20 years.

In this report, the Committee on Human and Environmental Exposure Science in the 21st Century presents a conceptual framework for exposure science and a vision for advancing exposure science in the 21st century. The committee describes scientific and technologic advances needed to support the vision and concludes with a discussion of the elements needed to realize it, including research and tool development, transagency coordination, education, and engagement of a broader stakeholder community.

This report has been reviewed in draft form by persons chosen for their diverse perspectives and technical expertise in accordance with procedures approved by the National Research Council Report Review Committee. The purpose of the independent review is to provide candid and critical comments that will assist the institution in making its published report as sound as possible and to ensure that the report meets institutional standards of objectivity, evidence, and responsiveness to the study charge. The review comments and draft manuscript remain confidential to protect the integrity of the deliberative process. We thank the following for their review of this report: Philip Landrigan, Mount Sinai School of Medicine; Jonathan Levy, Boston University School of Public Health; Rachel Morello-Frosch, University of California, Berkeley; Michael Newman, College of William & Mary; John Nuckols, JRN & Associates Environmental Health Sciences; Sean Philpott, Union Graduate College; Stephen Rappaport, University of California, Berkeley; Lawrence Reiter, U.S. Environmental Protection Agency (retired); Joyce Tsuji, Exponent; Mark Utell, University of Rochester School of Medicine and Dentistry; Craig Williamson, Miami University; Edward Zellers, University of Michigan.

Although the reviewers listed above have provided many constructive comments and suggestions, they were not asked to endorse the conclusions or recommendations, nor did they see the final draft of the report before its release. The review of the report was overseen by the review coordinator, Joseph V. Rodricks, ENVIRON, and the review monitor, Michael F. Goodchild, University of California, Santa Barbara. Appointed by the National Research Council, they were responsible for making certain that an independent examination of the report was carried out in accordance with institutional procedures and that all review comments were carefully considered. Responsibility for the final content of the report rests entirely with the committee and the institution.

The committee gratefully acknowledges the following for making presentations to the committee: Steven Bradbury, Helen Dawson, Sumit Gangwal, Elaine Cohen Hubal, Bryan Hubbell, Edward Ohanian, Lawrence Reiter (retired), Rita Schoeny, and Linda Sheldon, U.S. Environmental Protection Agency; Harry Cullings, Radiation Effects Research Foundation; Michael Dellarco, National Institute of Child Health and Human Development; Otto Hänninen and Matti Jantunen, Finland National Institute of Health and Welfare; Aubrey Miller, National Institute of Environmental Health Sciences; Chris Portier, Centers for Disease Control and Prevention; and Craig Postlewaite, U.S. Department of Defense.

The committee is also grateful for the assistance of National Research Council staff in preparing this report. Staff members who contributed to the effort are Eileen Abt, project director; James Reisa, director, Board on Environmental Studies and Toxicology; Keegan Sawyer, program officer; Keri Schaffer, research associate; Norman Grossblatt, senior editor; Mirsada Karalic-Loncarevic, manager, Technical Information Center; Radiah Rose, manager, editorial projects; Orin Luke, senior program assistant; and Tamara Dawson, program associate.

We especially thank the members of the committee for their efforts throughout the development of this report.

>Kirk R. Smith, *Chair*
>Paul J. Lioy, *Vice Chair*
>Committee on Human and
>Environmental Exposure
>Science in the 21st Century

Contents

SUMMARY 3

1 **INTRODUCTION** 19
   Background, 19
   Defining the Scope of Exposure Science, 22
   The Past Millennia, 24
   Opportunities and Challenges: The New Millennium, 28
   Roadmap, 31
   References, 35

2 **A VISION FOR EXPOSURE SCIENCE IN THE 21st CENTURY**............42
   References, 48

3 **APPLICATIONS OF EXPOSURE SCIENCE** 50
   Introduction, 50
   Epidemiology, 50
   Toxicology, 57
   Environmental Regulation, 60
   Environmental Planning, 67
   Disaster Management, 75
   Conclusions, 76
   References, 79

4 **DEMANDS FOR EXPOSURE SCIENCE**..90
   Introduction, 90
   Health and Environmental Science Demands, 92
   Market Demands, 97
   Societal Demands, 99
   Policy and Regulatory Demands, 100
   Building Capacity to Meet Demands, 101
   References, 101

5 SCIENTIFIC AND TECHNOLOGIC ADVANCES 106
Introduction, 106
Tracking Sources, Concentrations, and Receptors with
  Geographic Information Technologies, 108
Ubiquitous Sensing For Individual and Ecologic Exposure Assessment, 117
Biomonitoring for Assessing Internal Exposures, 128
Models, Knowledge, and Decisions, 134
References, 140

6 PROMOTING AND SUSTAINING PUBLIC TRUST
IN EXPOSURE SCIENCE 154
Protecting Research Volunteers, 155
Promoting Public Trust, 157
Community Engagement and Stakeholder Participation, 157
Use of Community-Based Participatory Research, 158
Challenges Ahead, 161
Guiding Values: The Right to Learn, 163
Conclusions, 165
References, 166

7 REALIZING THE VISION 169
Introduction, 169
The Exposure Data Landscape, 171
Immediate Challenges: Chemical Evaluation and Risk Assessment, 174
Implementing the Vision, 176
Research Needs, 176
Transagency Coordination, 179
Enabling Resources, 179
Conclusions, 181
References, 182

APPENDIXES

A BIOGRAPHIC INFORMATION ON THE COMMITTEE ON
   HUMAN AND ENVIRONMENTAL EXPOSURE SCIENCE IN
   THE 21st CENTURY 184

B STATEMENT OF TASK..191

C CONCEPTS AND TERMINOLOGY IN EXPOSURE SCIENCE 193

BOXES, FIGURES, AND TABLES

BOXES

1-1 Definition and Scope of Exposure Science, 20

1-2 Illustrations Demonstrating How the Degradation of the Ecosystems Due to Human Activities Increases Exposures to Chemical and Biologic Stressors, 33
3-1 Case Study of Exposure Assessment for the National Children's Study, 53
3-2 Case Study of the Hanford Environmental Dose-Reconstruction Project, 54
3-3 An Environment-Wide Association Study, 56
3-4 Value of Improved Exposure Estimates for Epidemiologic Studies, 57
3-5 Case Study of Perchlorate in Drinking Water, 63
3-6 Case Study of Chemicals in Breast Milk: Policy Action Based on Exposure Data, 66
3-7 Health Impact Assessment of Mobile Sources in San Francisco, 68
3-8 Exposure to Multiple Stressors in a Large Lake Ecosystem, 72
3-9 Emergency Management After the Attack on the World Trade Center, 77
5-1 Evaluating the Reliability of Aerosol Optical Depth Against Ground Observations, 110
5-2 Evaluation of MODIS 1 km Product, 110
5-3 Embedded Sensing of Traffic in Rome, 119
5-4 Ubiquitous Sensing of Physical Activity and Location, 119
5-5 Participatory Sensing, 121
5-6 Potential Application of –omics and Exposure Data in Personalized Medicine, 130
5-7 Global-Scale and Regional-Scale Models Used to Assess Human and Ecologic Exposure Potential in Terms of Long-Range Transport Potential and Persistence, 136
6-1 Case Study of Exposure Justice and Community Engagement: ReGenesis in Spartanburg, SC, 159

FIGURES

S-1 Conceptual framework showing the core elements of exposure science as related to humans and ecosystems, 6
S-2 Selected scientific and technologic advances for measuring and monitoring considered in relation to the conceptual framework shown in Figure S-1, 8
1-1 The classic environmental-health continuum, 21
1-2 Core elements of exposure science, 24
1-3 An illustration of how exposures can be measured or modeled at different levels of integration in space and time, from source to dose, and among different human, biologic, and geographic systems, 25
1-4 Connections between ecosystem services and human-well being, 32
3-1 General schema of exposure assessment in environmental epidemiology, 51
3-2 Exposure to Multiple Stressors in Lake Tahoe, 74
4-1 The four major demands for exposure science, 92
5-1 Selected scientific and technologic advances considered in relation to the conceptual framework, 109
5-2 Aerosol optical depth derived from MODIS data for the New England region, 111

5-3 Example of a binary buffer overlay showing people likely to experience traffic-related air-pollution exposure, 115
5-4 Map of a flood plain in the Netherlands showing secondary risk of poisoning by cadmium in Little Owls developed using a combination of measured cadmium concentrations, food web modeling, knowledge of foraging in different habitats, and probabilistic risk assessment, 116
5-5 Output from a CalFit telephone showing the location and activity level of volunteers in kilocalories per 10-second period in a pilot study in Barcelona, Spain, 120
C-1 Another view of the source-to-outcome continuum for exposure science, 194
C-2 Core elements of exposure science, 195

TABLES

5-1 Available Methods and Their Utility for Ecologic Exposure Assessment, 133

EXPOSURE SCIENCE
in the 21st Century

A VISION AND A STRATEGY

Summary

We are exposed every day to agents that have the potential to affect our health—through the personal products we use, the water we drink, the food we eat, the soil and surfaces we touch, and the air we breathe. Exposure science addresses the intensity and duration of contact of humans or other organisms with those agents (defined as chemical, physical, or biologic stressors)[1] and their fate in living systems. Exposure assessment, an application of this field of science, has been instrumental in helping to forecast, prevent, and mitigate exposures that lead to adverse human health or ecologic outcomes; to identify populations that have high exposures; to assess and manage human health and ecosystem risks; and to protect vulnerable and susceptible populations.

Exposure science has applications in public health and ecosystem protection, and in commercial, military, and policy contexts. It is central to tracking chemicals and other stressors that are introduced into global commerce and the environment at increasing rates, often with little information on their hazard potential. Exposure science is increasingly used in homeland security and in the protection of deployed soldiers. Rapid detection of potentially harmful radiation or hazardous chemicals is essential for protecting troops and the general public. The ability to detect chemical contaminants in drinking water at low but biologically relevant concentrations quickly can help to identify emerging health threats, and monitoring of harmful algal blooms and airborne pollen can help to identify health-relevant effects of a changing climate. With regard to policy and regulatory decisions, exposure information is critical in budget-constrained times for assessing the value of proposed public-health actions.

Exposure science has a long history, having evolved from such disciplines as industrial hygiene, radiation protection, and environmental toxicology into a theoretical and practical science that includes development of mathematical models and other tools for examining how individuals and populations come into contact with environmental stressors. Exposure science has played a fundamental role in the development and application of many fields related to

---

[1]Examples include chemical (toluene), biologic (*Mycobacterium tuberculosis*), and physical (noise) stressors.

environmental health, including toxicology, epidemiology, and risk assessment. For example, exposure information is critical in the design and interpretation of toxicology studies and is needed in epidemiology studies to compare outcomes in populations that have different exposure levels. Collection of better exposure data can provide more precise information regarding risk estimates and lead to improved public-health and ecosystem protection. For example, exposure science can improve characterization of populationwide exposure distributions, aggregate and cumulative exposures, and high-risk populations. Advancing and promoting exposure science will allow it to have a more effective role in toxicology, epidemiology, and risk assessment and to meet growing needs in environmental regulation, urban and ecosystem planning, and disaster management.

The committee identified emerging needs for exposure information. A central example is the knowledge gap resulting from the introduction of thousands of new chemicals into the market each year. Another example is the increasing need to address health effects of low-level exposures to chemical, biologic, and physical stressors over years or decades. Market demands also require the identification and control of exposures resulting from the manufacture, distribution, and sale of products. Societal demands for exposure data arise from the aspirations of individuals and communities—relying on an array of health, safety, and sustainability information—for example, to maintain local environments, personal health, the health of workers, and the global environment.

Recently, a number of activities have highlighted new opportunities for exposure science. For example, increasing collection and evaluation of biomarker data through the Centers for Disease Control and Prevention National Health and Nutrition Examination Survey and other government efforts offer a potential for improving the evaluation of source–exposure and exposure–disease relationships. The development of the "exposome", which conceptualizes that the totality of environmental exposures (including such factors as diet, stress, drug use, and infection) throughout a person's life can be identified, offers an intriguing direction for exposure science. And the publication of two recent National Research Council reports—*Toxicity Testing in the 21st Century: A Vision and a Strategy* (2007) and *Science and Decisions: Advancing Risk Assessment* (2009)—have substantially advanced conceptual and experimental approaches in companion fields of toxicology and risk assessment while presenting tremendous opportunities for the growth and development of exposure science.

The above activities have been made possible largely by advances in tools and technologies—sensor systems, analytic methods, molecular technologies, computational tools, and bioinformatics—over the last decade, which are providing the potential for exposure data to be more accurate and more comprehensive than was possible in the past. The scientific and technologic advances also provide the potential for the development of an integrated systems approach to exposure science that is more fully coordinated with other fields of environmental health; can address scientific, regulatory, and societal challenges better; can provide exposure information to a larger swath of the population; and can

Summary

embrace both human health and ecosystem protection. The availability of the massive quantities of individualized exposure data that will be generated might create ethical challenges and raise issues of privacy protection.

Recognizing the challenges and the need for a prospective examination of exposure science, the U.S. Environmental Protection Agency (EPA) and the National Institute of Environmental Health Sciences (NIEHS) asked the National Research Council to develop a long-range vision and a strategy for implementing the vision over the next 20 years, including development of a unifying conceptual framework for the advancement of exposure science.[2] In response to the request, the National Research Council convened the Committee on Human and Environmental Exposure Science in the 21st Century, which prepared this report.

In this summary, the committee presents a roadmap of how technologic innovations and strategic collaborations can move exposure science into the future. It begins with a discussion of a new conceptual framework for exposure science that is broadly applicable and relevant to all exposure media and routes, reflecting the current and expected needs of the field. It then describes scientific and technologic advances in exposure science. The committee next presents its vision for advancing exposure science in the 21st century. Finally, it discusses more broadly the elements needed to realize the vision, including research and tool development, transagency coordination, education, and engagement of a broad stakeholder community that includes government, industry, nongovernment organizations, and communities.

CONCEPTUAL FRAMEWORK

Exposure science can be thought of most simply as the study of stressors, receptors, and their contacts in the context of space and time. For example, ecosystems are receptors for such stressors as mercury, which may cascade from the ecosystem to populations to individuals in the ecosystem because of concentration and accumulation in the food web, which lead to exposure of humans and other species. As the stressor (mercury in this case) is absorbed into the bodies of organisms, it comes into contact with tissues and organs. It is important to recognize that exposure science applies to any level of biologic organization—ecologic, community, or individual—and, at the individual level, encompasses external exposure (outside the person or organism), internal exposure (inside the person or organism), and dose.

To illustrate the scope of exposure science and to embrace a broader view of the role that it plays in human health and ecosystem protection, the committee developed the conceptual framework shown in Figure S-1.

---

[2]Given the committee's statement of task, it addressed primarily exposure-science issues related to the U.S. and other developed countries.

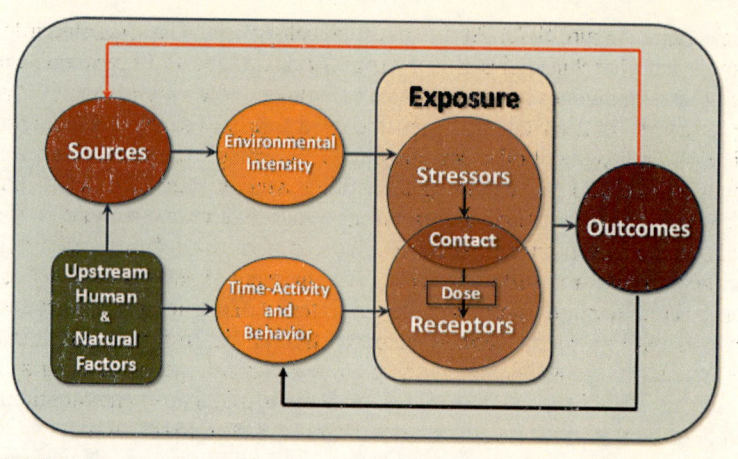

FIGURE S-1 Conceptual framework showing the core elements of exposure science as related to humans and ecosystems.

Figure S-1 identifies and links the core elements of exposure science: sources of stressors, environmental intensity[3] (such as pollutant concentrations), time–activity and behavior, contact of stressors and receptors, and outcomes of the contact. The figure shows the role of upstream human and natural factors in determining which stressors are mobilized and transported to key receptors. (Examples of those factors are choosing whether to use natural gas or diesel buses and choosing whether to pay more for gasoline and drive a car or to take a bus—the choices influence the sources and can influence behavior.) The figure indicates the role of the behavior of receptors and time in modifying contact, depending on environmental intensities that influence exposure. Figure S-1 encapsulates both external and internal environments within the "exposure" box, but indicates that exposure is measured at some boundary between source and receptor. Dose is the amount of material that passes or otherwise has influence across the boundary; comes into contact with the target system, organ, or cell; and produces an outcome. For example, a dose in one tissue, such as the blood, can serve as the exposure of another tissue that the blood perfuses.

SCIENTIFIC AND TECHNOLOGIC ADVANCES

Innovations in science and technology enable advances to be made in exposure science. Numerous state-of-the-art methods and technologies measure exposures, from external concentrations to personal exposures to internal exposures. (Selected technologies considered in relation to the conceptual framework

---

[3]*Intensity* is the preferred term because some stressors, such as temperature excesses, cannot be easily measured as concentrations.

are included in Figure S-2.) For example, developments in geographic information science and technologies are leading to rapid adoption of new information from satellites via remote sensing and providing immediate access to data on potential environmental threats. Improved information on physical activity and locations of humans and other species obtained with global positioning systems and related geolocation technologies is increasingly combined with cellular-telephone technologies. Biologic monitoring and sensing increasingly offer the potential to assess internal exposures. In addition, models and information-management tools are needed to manage the massive quantities of data that will be generated and to interpret the complex interactions among receptors and environmental stressors. The convergence of those scientific methods and technologies raises the possibility that in the near future integrated sensing systems will facilitate individual-level exposure assessments in large populations of humans or other species. The various technologies are discussed below.

Tracking Sources, Concentrations, and Receptors with Geographic Information Technologies

Geographic information technologies—remote sensing, global positioning and related locational technologies, and geographic information systems (GIS)—are motivating an emphasis on spatial information in exposure science. They can be used to characterize sources and concentrations and can improve understanding of stressors and receptors when used in concert with other methods and data.

- *Remote sensing* involves the capture, retrieval, analysis, and display of information on subsurface, surface, and atmospheric conditions that is collected by using satellite, aircraft, or other technologies. Remote sensing is an important method for improving our capacity to assess human and ecologic exposures as it provides global information on the earth's surface, water, and atmosphere, and it can provide exposure estimates in regions where available ground observation systems are sparse. For example, data collected with remote sensing over "Ground Zero" was used initially to assess the potential asbestos hazards related to the dust that settled over lower Manhattan after the collapse of the World Trade Center towers. Remote sensing of vegetation combined with GIS has been used to assess potential exposure of wildlife to pesticides and metals.
- *Global positioning system (GPS) and geolocation technologies*—which are now embedded into many cellular telephones, vehicle navigation systems, and other instruments—provide a means of tracking the geographic position of a person or other species. Geolocation technologies have been used extensively in exposure-assessment studies, are important for providing accurate information on the location of an individual or species in space and time, and offer precise

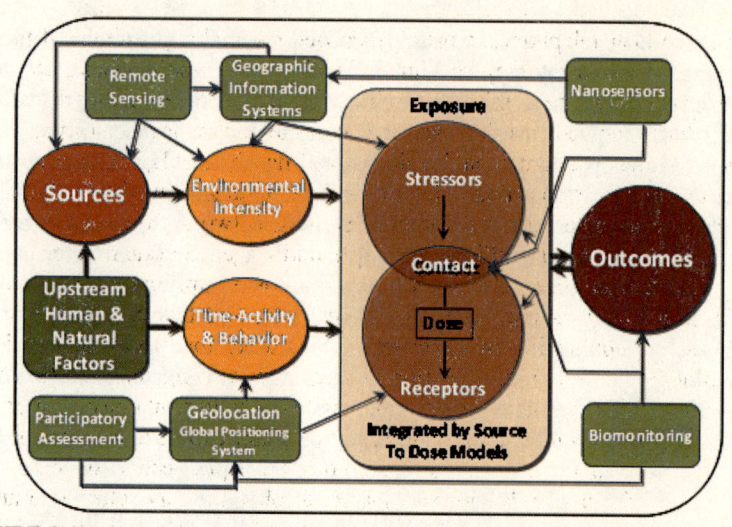

FIGURE S-2 Selected scientific and technologic advances for measuring and monitoring considered in relation to the conceptual framework shown in Figure S-1.

exposure estimates. When geolocation data (with information on air or water quality) are used with activity measurements readily available through portable accelerometers, additional information can be inferred about potential uptake of stressors.

- *GIS* allows storage and integration of data from different sources (for example, exposure information and health characteristics of populations) by geographic location. It also provides quantitative information on the topologic relationship between an exposure source and a receptor, which allows researchers to characterize proximity to roadways, factories, water bodies, and other land uses. For example, GIS used with modeling data has provided information on exposure exceedances of threatened and endangered species associated with environmental contaminants. Web-based GIS increasingly serves as a tool for educating and empowering communities to understand and manage environmental exposures.

The increasing use of geographic information technologies (for example, through cellular telephones, GPS, or Web-based systems), many of which are operated by the private sector, raises important issues about privacy protection and the use of the resulting data by exposure-science researchers for improving public health.

Ubiquitous Sensing

Over the last 20 years, there have been substantial advances in personal environmental-monitoring technologies. The advances have been made possible in part by cellular telephones, which are carried routinely by billions of people throughout the world and may be equipped with motion, audio, visual, and location sensors that can be manipulated with cellular or wireless networks. Pollution-monitoring devices can be integrated into the telephones (for example, for measuring particulate matter and volatile organic chemicals). In this context, cellular telephones, supporting software, and expanding networks (cellular and WiFi) can be used to form "ubiquitous" sensing networks to collect personal exposure information on millions of people and large ecosystems. People can then act as "citizen–scientists", collecting their own exposure data to inform themselves about what they might be exposed to, and this can lead to more comprehensive application of exposure-science tools for health and environmental protection. However, validation of ubiquitous sensing networks to ensure the accuracy and precision of the data collected is an important consideration.

Developing ubiquitous monitoring for personal exposure assessment will depend on rapid advances in sensor technologies. Despite recent advances, personal sensors still have only modest capacity to obtain highly selective, multistressor measurements. There is a need for a wearable sensor that is capable of monitoring multiple analytes in real time. Such a device would allow more rapid identification of "highly exposed" people to help to identify sources and means of reducing exposures. Recent advances in nanoscience and in nanotechnology offer an unprecedented opportunity to develop very small, integrated sensors that can overcome current limitations.

With regard specifically to environmental exposure, advances in electronic miniaturization of sensors and data management are motivating the development of environmental sensor networks that can provide long-term real-time exposure-monitoring data on our ecosystem. Much of the interest in network sensors has been motivated by national-security concerns, including concerns about monitoring drinking-water or air quality.

Biomonitoring for Assessing Internal Exposures

With advances in genomic techniques and informatics, exposure science is moving from collection of external exposure information on a small number of stressors, locations, times, and individuals to a more systematic assemblage of internal exposures to multiple stressors in individuals in human populations and multiple species in our environment.

The committee considered three broad topics in biomonitoring: measures of internal exposure, biosignatures of exposure, and measurement of biochemical modifiers of internal exposure.

- *Measures of internal exposure* to stressors are closer to the target site of action for biologic effects than are external measures of exposure and so improve the correlation of exposure with effects. Analytic methods enable the detection of low concentrations of multiple stressors. The measurement of thousands of small organic molecules in biologic samples with metabolomics is now being applied to biomonitoring of chemicals in humans and in wildlife. Such approaches are not limited to a chemical or class of chemicals selected in advance but rather provide broader, agnostic assessment that can identify exposures and potentially improve surveillance and elucidate emerging stressors. Proteomics and adductomics expand the types of internal measures of exposure that can be analyzed, including the analysis of compounds in the blood that have short half-lives, such as oxidants in cigarette smoke and acrylamide. Rapidly evolving sensor platforms linked to physiologically based pharmacokinetic (PBPK)[4] models are expected to enable field measurements of chemical samples in blood, urine, or saliva from human and nonhuman populations and rapid interpretation of the concentrations in the samples. However, inferring the sources and routes of these internal exposures remains a research challenge.

- *Biosignatures of exposure* reflect the net biologic effect of internal exposure to stressors that act on specific biologic pathways. For example, oxidative modifications of DNA or protein can be used to represent the net internal exposure to oxidants and antioxidants. Biosignatures provide better assessment of exposure–disease correlations, but they are still limited in their ability to target reduction in any specific compound or source.

- *Measurement of biochemical modifiers of internal exposure* can be used qualitatively to identify populations that are expected to have greater internal exposures to a given stressor (for example, because of differences in metabolism or higher absorption) or quantitatively by inclusion in PBPK–pharmacodynamic models used for exposure assessment and prediction of doses. Transcriptomics, proteomics, and to a smaller extent metabolomics offer the ability to measure the status of key biologic processes that affect the pharmacokinetics (that is the absorption, distribution, metabolism, or elimination) of chemical stressors.

With regard to ecologic exposure assessment, the use of molecular techniques as biomarkers to assess ecologic exposure to stressors is limited in that most of these techniques cannot be linked quantitatively to the level of exposure and are not highly selective. There is a need to develop rapid-response, quantitative exposure-assessment tools that can provide useful information for exposure assessment in ecologic risk assessments.

Models and Information-Management Tools

Models and information-management tools are critical for interpreting and managing the quantities of data being generated with the expanding technologies. For example, satellite imaging and personal monitoring techniques are

---

[4] A mathematical modeling technique for predicting the absorption, distribution, metabolism, and excretion of a compound in humans or other animal species.

generating enormous quantities of spatiotemporal data and information on people's movements and activities, and biologic assays are capable of monitoring millions of genetic variants, metabolites, or gene-expression or epigenetic changes in thousands of subjects. The ability of models to provide a repository for exposure information, to help in interpreting data and observations, and to provide tools for predicting trends will continue to be a cornerstone of exposure science.

Many types of models will continue to be important in exposure science—for example, activity-based models for tracking the history of individuals or populations and process-based models for tracking the movement of stressors from source to receptor—but there is a growing need for structure–activity models that can classify chemicals with regard to exposure and potential health effects.

The key to the future of exposure models is how they incorporate the increasing number of observations that are being collected. Although observations alone are important, it is their analysis, through application of models, that elucidates the value of the measurements. It is also important to quantify the uncertainty in the exposure estimates provided by models. However, to fully address environmental health concerns, exposure models need to be systematically integrated into source to dose modeling systems.

Informatics encompasses tools for managing, exploring, and integrating massive amounts of information from diverse sources and in widely different formats. Informatics relies on model algorithms, databases and information systems, and Web technologies. Although it is highly developed in biology and medicine, its application in exposure science is in its infancy; informatics offers great promise for improving the linkages of exposure science to related environmental-health fields.

A number of informatics efforts are under way. For example, ExpoCast Database, developed as part of EPA's Expocast program to advance the characterization of exposure to address the new toxicity-testing paradigm, is designed to house measurements from human exposure studies and to support standardized reporting of observational exposure information. Recently, a pilot Environment-Wide Association Study was conducted in which exposure–biomarker and disease-status data were systematically interpreted in a manner analogous to that in a Genome-Wide Association Study.[5] In addition, the exposure field has developed and designed an exposure ontology[6] to facilitate centralization and integration of exposure data with data in other fields of environmental health, including toxicology, epidemiology, and disease surveillance.

---

[5]Genome-Wide Association Studies are epidemiologic studies that examine the associations between particular genetic variants and specific disease outcomes.

[6]Ontologies, specifications of the terms and their logical relationships used in a particular field, are used to improve search capabilities and allow mapping of relationships among different databases and informatics systems.

A VISION FOR EXPOSURE SCIENCE IN THE 21st CENTURY

New challenges and new scientific advances impel us to an expanded vision of exposure science. The vision is intended to move the field from its historical origins—where it has typically addressed discrete exposures with a focus on either external or internal environments and a focus on either effects of sources or effects on biologic systems, one stressor at a time—to an integrated approach that considers exposures from source to dose, on multiple levels of integration (including time, space, and biologic scale), to multiple stressors, and scaled from molecular systems to individuals, populations, and ecosystems.

The vision, the "eco-exposome", is defined as the extension of exposure science from the point of contact between stressor and receptor inward into the organism and outward to the general environment, including the ecosphere. Adoption and validation of the eco-exposome concept should lead to the development of a universal exposure-tracking framework that allows the creation of an exposure narrative, the prediction of biologically relevant human and ecologic exposures, and the generation of improved exposure information for making informed decisions on human and ecosystem health protection. The vision is premised on the scientific developments of the last decade.

To advance this broader vision, exposure science needs to deliver knowledge that is effective, timely, and relevant to current and future environmental-health challenges. To do so, exposure science needs to continue to build capacity to

- *Assess and mitigate exposures quickly in the face of emerging environmental-health threats and natural and human-caused disasters.* For example, this requires expanding techniques for rapid measurement of single and multiple stressors on diverse geographic, temporal, and biologic scales. That includes developing more portable instruments and new techniques in biologic and environmental monitoring to enable faster identification of chemical, biologic, and physical stressors that are affecting humans or ecosystems.

- *Predict and anticipate human and ecologic exposures related to existing and emerging threats.* Development of models or modeling systems will enable us to anticipate exposures and characterize exposures that had not been previously considered. For example, predictive tools will enable development of exposure information on thousands of chemicals that are now in widespread use and enable informed safety assessments of existing and new applications for them. In addition, strategic use of such diverse information as structural properties of chemicals, nontargeted environmental surveillance, biomonitoring, and modeling tools are needed for identification and quantification of relevant exposures that may pose a threat to ecosystems or human health.

- *Customize solutions that are scaled to identified problems.* As stated in *Science and Decisions: Advancing Risk Assessment* (2009), the first step in a risk assessment should involve defining the scope of the assessment in the context of the decision that needs to be made. Adaptive exposure assessments could facilitate that approach by tailoring the level of detail to the problem that needs to be addressed. Such an assessment may take various forms, including very narrowly focused studies, assessments that evaluate exposures to multiple stressors to facilitate cumulative risk assessment, or assessments that focus on vulnerable or susceptible populations.

- *Engage stakeholders associated with the development, review, and use of exposure-science information, including regulatory and health agencies and groups that might be disproportionately affected by exposures*—that is, engage broader audiences in ways that contribute to problem formulation, monitoring and data collection, access to data, and development of decision-making tools. Ultimately, the scientific results derived from the research will empower individuals, communities, and agencies to prevent and reduce exposures and to address environmental disparities.

For the committee's vision to be realized in light of resource constraints, priorities need to be set among research and resource needs that focus on the problems or issues that are critically important for addressing human and ecologic health. For example, screening-level exposure information may be adequate to address some questions, targeted data may be useful for others, and extensive data may be required in some circumstances. Health-protective default assumptions can provide incentives for data generation and can allow timely decisions despite inevitable data gaps.

REALIZING THE VISION

The demand for exposure information, coupled with the development of tools and approaches for collecting and analyzing such data, has created an opportunity to transform exposure science to advance human and ecosystem health. The transformation will require an investment of resources and a substantial shift in how exposure-science research is conducted and its results implemented. In the near term exposure science needs to develop strategies to expand exposure information rapidly to improve our understanding of where, when, and how exposures occur and their health significance. Data generated and collected should be used to evaluate and improve models for use in generating hypotheses and developing policies. New exposure infrastructure (for example, sensor networks, environmental monitoring, activity tracking, and data storage and distribution systems) will help to identify the largest knowledge gaps and reveal where gathering more exposure information would contribute the most to reducing uncertainty.

For example, more exposure data needs to be collected to populate emerging exposure databases (for example EPA's ExpoCast Database) and to develop tools to systematically mine various sources of exposure information, so as to bridge the gap between exposure science and other environmental-health disciplines. New and improved surveillance systems can be designed to increase our knowledge of environmental stressors and provide information for estimation and characterization of exposures. Targeted exposure studies will be essential for gathering detailed information on hot spots or places of highest potential impact to vulnerable and susceptible populations. Surveillance programs together with targeted exposure-measurement programs will help build a predictive exposure network that can address environmental-health questions.

Research Needs

To implement its vision, the committee identified research needs that call for further development of existing and emerging methods and approaches, validation of methods and their enhancement for application on different scales and in broader circumstances, and improved linkages to research in other sectors of the environmental-health sciences. The research needs are organized into several broad categories addressed below, and they are organized by priority within each category on the basis of the time that will be required for their development and implementation: *short term* denotes less than 5 years, *intermediate term* 5-10 years, and *long term* 10-20 years.

Providing effective responses to immediate or short-term public-health or ecologic risks requires research on observational methods, data management, and models:

Short term

- Identify, improve, and test instruments that can provide real-time tracking of biologic, chemical, and physical stressors to monitor community and occupational exposures to multiple stressors during natural, accidental, or terrorist events or during combat and acts of war.
- Explore, evaluate, and promote the types of targeted population-based exposure studies that can provide information needed to infer the time course of internal and external exposures to high-priority chemicals.

Intermediate term

- Develop informatics technologies (software and hardware) that can transform exposure and environmental databases that address different levels of integration (time scales, geographic scales, and population types) into formats that can be easily and routinely linked with populationwide outcome databases (for humans and ecosystems) and linked to source-to-dose modeling platforms to facilitate rapid discovery of new hazards and to enhance preparedness and timely response.

Summary

- Identify, test, and deploy extant remote sensing, personal monitoring techniques, and source to dose model-integration tools that can quantify multiple routes of exposure (inhalation, ingestion, and dermal uptake) and obtain results that can, for example, be integrated with emerging methods (such as –omics technologies)[7] for tracking internal exposures.

Long term

- Enhance tracking of human exposures to pathogens on the basis of a holistic ecosystem perspective from source through receptor.

Supporting research on health and ecologic effects that addresses past, current, and emerging outcomes:

Short term

- Coordinate research with human-health and ecologic-health scientists to identify, collect, and evaluate data that capture internal and external markers of exposure in a format that improves the analysis and modeling of exposure–response relationships and links to high-throughput toxicity testing.
- Explore options for using data obtained on individuals and populations through market-based and product-use research to improve exposure information used in epidemiologic studies and in risk assessments.

Intermediate term

- Develop methods for addressing data and model uncertainty and evaluate model performance to achieve parsimony in describing and predicting the complex pathways that link sources and stressors to outcomes.
- Improve integration of information on human behavior and activities for predicting, mitigating, and preventing adverse exposures.

Long term

- Adapt hybrid designs for field studies to combine individual-level and group-level measurements for single and multiple routes of exposure to provide exposure data of greater resolution in space and time.

Addressing demands for exposure information among communities, governments, and industries with research that is focused, solution-based, and responsive to a broad array of audiences:

---

[7]Technologies used to identify and quantify all members of particular cellular constituents, for example, proteins (proteomics), metabolites (metabolomics), or lipids (lipomics).

Short term

- Develop methods to test consumer products and chemicals in premarketing controlled studies to identify stressors that have a high potential for exposure combined with a potential for toxicity to humans or ecologic receptors.
- Develop and evaluate cost-effective, standardized, non-targeted, and ubiquitous methods for obtaining exposure information to assess trends, disparities among populations (human and ecologic), geographic hot spots, cumulative exposures, and predictors of vulnerability.

Intermediate term

- Apply adaptive environmental-management approaches to understand the linkages between adverse exposures in humans and ecosystems better.
- Implement strategies to engage communities, particularly vulnerable or hot-spot communities, in a collaborative process to identify, evaluate, and mitigate exposures.

Long term

- Expand research in ways to use exposure science to more effectively regulate environmental risks in natural and human systems, including the built environment.

Transagency Coordination

Exposure science is relevant to the mission of many federal agencies, and transagency collaboration for exposure science in the 21st century would accelerate progress in and transform the field. Tox21—a collaboration among EPA, the National Institutes of Health (NIH), and the Food and Drug Administration—that was established to leverage resources to advance the recommendations in the 2007 National Research Council report *Toxicity Testing in the 21st Century: A Vision and a* Strategy serves as a relevant model. The present committee considers that the model used in establishing Tox21 could be extended to exposure science and lead to the creation of Exposure21. Exposure21, in addition to engaging the stakeholders (government, industry, and nongovernment organizations) involved in Tox21, would need to be extended to other federal agencies—such as the U.S. Geological Survey, the Centers for Disease Control and Prevention, the National Oceanic and Atmospheric Administration, the National Science Foundation (NSF), and the National Aeronautics and Space Administration—to promote greater access to and sharing of data and resources on a broader scale. Including them would provide access to resources for transformative technology innovations, for example, in nanosensors.

Enabling Resources

As the collaborative partnerships among agencies are expanded, there will be opportunities to share research results, to demonstrate the value of exposure-science research to other agencies and decision-makers, and to generate additional resources. The committee recommends that intramural and extramural programs in EPA, NIEHS, the Department of Defense, and other agencies that advance exposure-science research be supported as the value of the research and the need for exposure information become more apparent.

Much of the human-based research in environmental-health sciences is funded by NIH. However, none of the existing study sections that review grant applications has substantial expertise in exposure science, and most study sections are organized around disease processes. In light of that and the role that an understanding of environmental exposures can play in disease prevention, a rethinking of how NIH study sections are organized that incorporates a greater focus on exposure science would allow a core group of experts to foster the objectives of exposure-science research. In addition, an increase in collaborations between agencies should be explored; for example, collaborations between EPA, NIEHS, and NSF could support integrative research between ecosystem and human-health approaches in exposure science. However, many other agencies engaged in exposure-science research could be included in the collaborations.

Because of the need to understand and prevent harmful exposures and risks in our society, EPA and NIEHS need to be able to work with the academic community to conduct exposure studies in all populations, particularly among the most vulnerable (for example, the elderly, children, and the infirm), under appropriate ethical guidelines.

The effective implementation of the committee's vision will depend on development and cultivation of scientists, engineers, and technical experts with experience in multiple fields to educate the next generation of exposure scientists and to provide opportunities for members of other fields to cross-train in the techniques and models used to analyze and collect exposure data. Exposure scientists will need the skills to collaborate closely with other fields of expertise, including engineers, epidemiologists, molecular and system biologists, clinicians, statisticians, and social scientists. To achieve that, the committee considers that the following are needed:

- An increase in the number of academic predoctoral and postdoctoral training programs in exposure science throughout the U.S. supported by training grants. NIEHS currently funds one training grant in exposure science; additional grants are needed.
- Short-term training and certification programs in exposure science for midcareer scientists in related fields.

- Development, by federal agencies that support human and environmental exposure science, of educational programs to improve public understanding of exposure-assessment research, including ethical considerations involved in the research. The programs would need to engage members of the general public, specialists in research oversight, and specific communities that are disproportionately exposed to environmental stressors.

Participatory and Community-Based Research Programs

To engage broader audiences, including the public, the committee suggests the development of more user-friendly and less expensive monitoring equipment to allow trained people in communities to collect and upload their own data in partnership with researchers. Such partnerships would improve the value of the data collected and make more data available for purposes of priority-setting and informing policy. One approach might include implementing a system of ubiquitous sensors (for example, through the use of cellular telephones, GPS, or other technologies) in two American cities to evaluate the feasibility of such systems to develop community-based exposure data that are reliable. Potential issues of privacy protection would need to be considered.

CONCLUDING REMARKS

Exposure information is crucial for predicting, preventing, and reducing human health and ecosystem risks. Exposure science has historically been limited by the availability of methods, technologies, and resources, but recent advances present an unprecedented opportunity to develop more rapid, cost-effective, and relevant exposure assessments. Research supported by such federal agencies as EPA and NIEHS has provided valuable partnership opportunities for building capacity to develop the technologies, resources, and educational structure that will be needed to achieve the committee's vision for exposure science in the 21st century.

1

Introduction

BACKGROUND

Exposure science is essential for the protection of public health and the environment. However, the challenges and opportunities for exposure science are considerable. The ability to address them will influence advances in human health and ecosystem protection. Exposure science also will play a role in decision-making in other arenas, including consumer-product safety, environmental planning, climate-change mitigation, and energy development. This report provides a roadmap to navigate the future of exposure science to achieve greater integration and maximize its utility in the environmental and occupational health sciences, environmental-systems science, risk assessment, sustainability science, and industrial ecology.

Exposure science addresses the contact of humans and other organisms with chemical, physical, or biologic (CBP) stressors[1] (EPA 2003; EPA 2011b) over space and time and the fate of these stressors within the ecosystem and organisms—including humans. Although methods of assessment will depend on the situation, exposure science has two primary goals: to understand how stressors affect human and ecosystem health and to prevent or reduce contact with harmful stressors or to promote contact with beneficial stressors to improve public and ecosystem health. The impact of environmental stressors on human and ecologic health is enormous.

For example, the World Health Organization estimates that 24-40% of global disease burden (healthy life-years lost) can be attributed to environmental factors (Smith et al. 1999; WHO 2004; Prüss-Üstün and Corvalán 2006). However, it is not possible to be exact in such calculations, partly because what is "environmental" is not defined consistently (see Box 1-1 for use of the term *environmental* in this report). In a burden-of-disease context, environmental

---

[1]The Environmental Protection Agency defines a stressor as "any physical, chemical, or biological entity that can induce an adverse response" (EPA 2011a).

factors play a role in nearly all diseases, even ones that are not caused directly by environmental risk factors, by altering the course of disease initiated by other causes. In addition, if the total burden of disease is simply decomposed into "nature" or "nurture", it fails to account directly for the possibly large proportion that could be due to the interplay between the two (gene–environment interactions). Improving our understanding of environmental factors and their relationships with disease is critical for preventing illness and death.

With respect to ecosystems, the 1999 National Research Council report *Our Common Journey: A Transition Toward Sustainability* (NRC1999) reported that the rising losses of wild nature, species number, species diversity, and ecosystem integrity were associated with exposures to environmental stressors, including those related to urban and agricultural land conversion and climate change.

Figure 1-1 illustrates the relationship of exposure to other key elements along the environmental-health continuum from the source of a stressor to an outcome. This figure has evolved from previous diagrams (for example, Smith 1988a; Lioy 1990; NRC 1998; EPA 2009a). For more than 20 years, this framework has demonstrated the central role of exposure science in environmental health science in that exposure sits midway between the sources of pollution (and other stressors) on the left—elements that typically can be controlled—and adverse health outcomes on the right, which need to be prevented. Exposure is strategically located upstream of dose and yet provides information and metrics that inform source control and health risk.

BOX 1-1 Definition and Scope of Exposure Science

Exposure science is defined by this committee as the collection and analysis of quantitative and qualitative information needed to understand the nature of contact between receptors (such as people or ecosystems) and physical, chemical, or biologic stressors. Exposure science strives to create a narrative that captures the spatial and temporal dimensions of exposure events with respect to acute and long-term effects on human populations and ecosystems.

For the purposes of this report, the committee focuses on environmental risk factors and excludes behavioral or lifestyle factors—such as diet, alcohol, and smoking—although it includes contaminants in food, water, and environmental tobacco smoke. It also excludes social risk factors (for example, crime and child abuse) but does consider them as modifying influences on exposures to stressors (Smith et al. 1999). The influence of social factors on environmental exposures is an area of active research. Natural hazards (for example, weather and arsenic contamination) are included here.

A central theme of this report is the interplay between the external and internal environments and the opportunity for exposure science to exploit novel technologies for assessing biologically active internal exposures from external sources.

Introduction

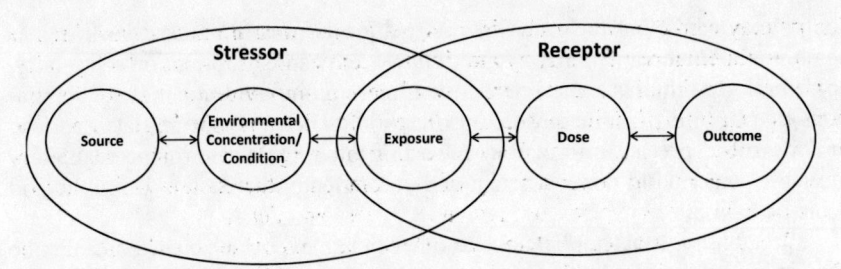

FIGURE 1-1 The classic environmental-health continuum. Figure 1-2 illustrates the revised version discussed in the present report. Source: Adapted from EPA 2009a.

There are many notable examples of the roles that exposure science can play in protecting public health. Consider how measurements of childhood blood lead concentrations since the 1970s reveal the dramatic efficacy of lead removal from gasoline in reducing exposure to this neurotoxicant in children (Muntner et al. 2005; Jones et al. 2009). Population-scale measurements of cotinine in urine document the reduction of exposure to second-hand tobacco smoke that resulted from control of tobacco-smoking in the workplace and public areas (EPA SAB 1992). Exposure modeling from the U.S. Environmental Protection Agency's (EPA) National-Scale Air Toxics Assessment program has provided valuable information for communities on their exposure sources, concentrations, and risks and has helped to shed light on exposure disparities and environmental-justice issues (for example, Pastor et al. 2005).

Exposure science has played a critical role in understanding the influence of stressors on ecologic systems. For example, extensive exposure assessments of polycyclic aromatic hydrocarbons (PAHs) have been linked to liver damage in bottom-dwelling fish in Puget Sound, and field studies have demonstrated that containment of PAH sources has led to declines in PAH concentrations and a resulting decline in liver damage in fish (Myers et al. 2003).

Exposure science has applications in industrial, military, commercial, and global contexts. It is central to tracking chemicals and other agents that are introduced into global commerce at increasing rates, often with little information on their hazard potential (GAO 2005). Increasingly, exposure science is used for homeland security and the protection of deployed soldiers. Rapid detection of potentially harmful radiation or toxic chemicals is essential for protecting troops and the general public (IOM 2000). The ability to detect chemical contaminants in drinking water at low, biologically relevant concentrations quickly can help to identify emerging health threats, and monitoring of harmful algal blooms and airborne pollen can help to identify health-relevant effects of a changing climate.

As described in more detail in Chapter 3, applications of exposure science are critical for toxicology, epidemiology, risk assessment, and risk management. For example, toxicology provides information about how different chemical concentrations may affect public or ecologic health in laboratory studies or computer models, but the value of information is greatly increased when it is combined with comprehensive and reliable exposure information. Similarly, epidemiology requires exposure information to compare outcomes in populations that have different exposures. Collection of better exposure data can also provide more precise information regarding alternative control or regulatory measures and lead to more efficient and cost-effective protection of public and ecologic health.

In addition to its applications to other fields, exposure science data can be used independently to define trends, assess spatial or population variability, provide information on prevention and intervention, identify populations or ecosystems that have disproportionate exposures, and evaluate regulatory effectiveness.[2]

Exposure science is also poised to play a critical role in improving the ability to understand and address increasingly important human health and ecologic challenges and to support the development of sustainable industrial, agricultural, and energy technologies. Recognizing the need for a prospective examination of exposure science, EPA and the National Institute of Environmental Health Sciences asked the National Research Council to develop a long-range vision for exposure science and a strategy for implementing the vision over the next 20 years (see Appendix B for statement of task). In response to this request, the National Research Council convened the Committee on Human and Environmental Exposure Science in the 21st Century. The committee—which comprised experts in monitoring, modeling, environmental transport and transformation, geographic information science and related technologies, measurement and analytic techniques, risk assessment and risk management, epidemiology, occupational health, risk communication, ethics, informatics, and ecologic services—prepared this report.

DEFINING THE SCOPE OF EXPOSURE SCIENCE

Exposure science—sometimes defined as the study of the contact between receptors (such as humans or ecosystems) and physical, chemical, or biologic stressors—can be thought of most simply as the study of stressors, receptors, and their contact, including the roles of space and time. For example, ecosystems are receptors for such stressors as mercury, which may cascade from the ecosystem to populations to individuals within the ecosystem because of concentration and accumulation in the food web, which leads to exposure of humans and other species. As the stressor (mercury in this case) is absorbed into

---

[2]In 2011, the International Society of Exposure Science and the *Journal of Exposure Science and Environmental Epidemiology* published a compendium of digests (Graham 2011) that illustrate situations in which application of exposure science resulted in substantial health or policy benefits and situations in which lack of exposure information resulted in adverse consequences.

Introduction

the bodies of individuals, it may come into contact with other receptors, such as tissues and organs.

As the scientific communities generating and using exposure data have evolved, so have the terms and definitions used to characterize exposures. Some refer to dose (exposure dose, target dose, or external dose), others to exposure (for example, external or internal exposure), and yet others to an amalgam (exposure is external, dose is internal). A consistent language for the field of exposure science is important for communicating within the field and among disciplines and for developing exposure-science metrics for source monitoring and exposure prevention and reduction. The evolution of the field over the past 15 years has included a greater emphasis on the use of internal markers of exposure to assist in defining exposure-response relationships. As such, the conceptual basis of the field includes both external and internal exposures, using external measurement and modeling methods and internal markers as tools for characterizing past or current exposures. Appendix C provides more detailed discussion on the application of this terminology.

To reflect the definition of exposure science and to embrace a broader view of the role that exposure science plays in human-health and ecosystem-health protection, the committee developed the conceptual framework in Figure 1-2.

The conceptual framework identifies and links the core elements of exposure science: sources of stressors, environmental intensity (such as pollutant concentrations[3]), time–activity and behavior, contact of stressors and receptors, and outcomes of contact. Figure 1-2 shows the role of upstream human and natural factors in determining which stressors are mobilized and transported to key receptors. (Examples of factors include choosing to use natural gas vs diesel buses, or choosing to pay more for gasoline to drive a car vs taking the bus, where the choice influences the source and can also influence behavior.) It indicates the role of behavior of receptors and time in modifying the contact that results from environmental intensities that influence exposure. It brings both external and internal environments within exposure but retains the idea that exposure is measured at some boundary between the source and receptor and that dose is the amount of material that passes or otherwise has influence across the boundary to come into contact with the target system, organ, or cell and produces an outcome. For example, a dose in one tissue, such as the blood, can serve as an exposure of another tissue that the blood perfuses. Figure 1-2 recognizes the feedbacks inherent in exposure science. Consider, for example, how behavior changes in a diseased person or organism and influences exposure. The outcome can also affect the source, as when a person who has an environmentally mediated infectious disease becomes a source of pathogens in water supplies (Eisenberg et al. 2005).

---

[3] *Intensity* is the preferred term because some stressors, such as temperature excesses, cannot be easily measured as concentrations.

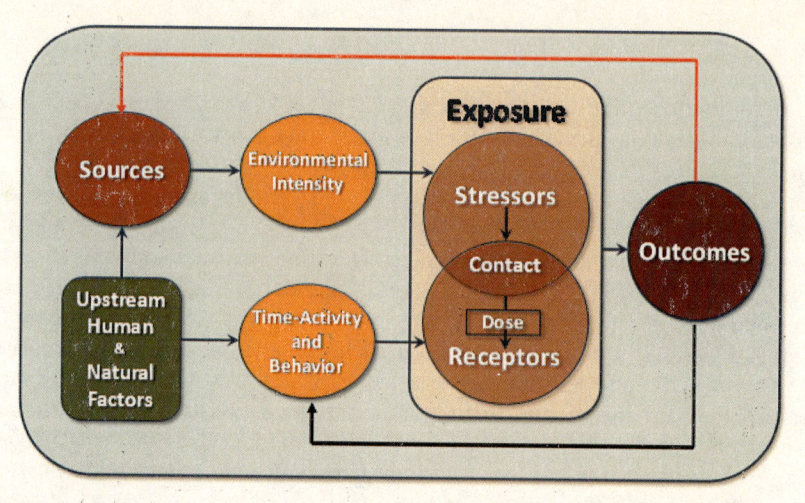

FIGURE 1-2 Core elements of exposure science.

Figure 1-3 frames an exposure narrative that plays out in space and time, and is intended to elucidate the stressor-receptor linkages at different levels of intergration. As a human (or fish, bird, or other organism) has changing contacts with different habitats, the intensity of a stressor changes, as do the number and duration of contacts. Here, exposure amounts to a multidimensional description of the location, time, and intensity of the target–stressor contacts. The exposure narrative covers relationships between receptors and locations and between locations and stressors; it provides a basis for drawing inferences about receptor–outcome relationships. That often requires recognition that any receptor can be associated with multiple environments (locations) and that locations can be associated with multiple stressors. Exposure science can be applied at any level of biologic organization—ecologic, community, or individual—and, within the individual, at the level of external exposure, internal exposure, or dose.

THE PAST MILLENNIA

To appreciate the vision for exposure science in the 21st century (discussed in Chapter 2), it is important to understand its historical context. Exposure science arose from such disciplines as industrial hygiene, radiation protection, and environmental toxicology, in which the importance of assessing exposure has been demonstrated. In one of the earliest efforts to address exposure, the ancient Greek physician Hippocrates (about 400 BC) demonstrated in his treatise *Air, Water, and Places* that the appearance of disease in human populations is influenced by the quality of air, water, and food; the topography of the land; and general living habits (Wasserstein 1982). In the 1500s, the

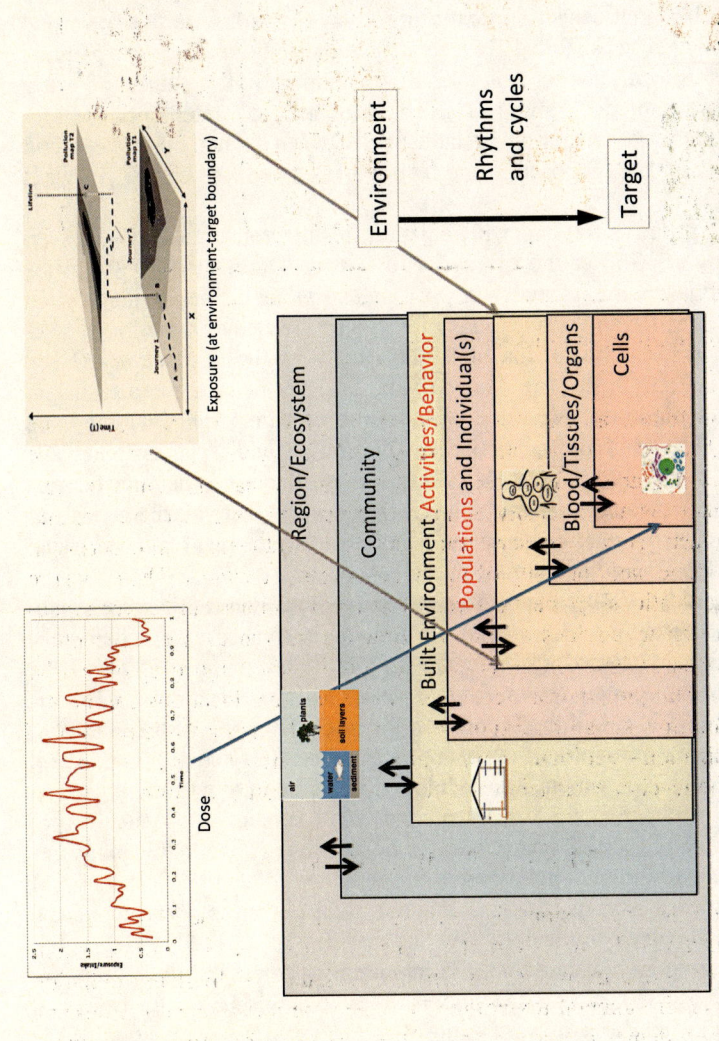

FIGURE 1-3 An illustration of how exposures can be measured or modeled at different levels of integration in space and time, from source to dose, and among different human, biologic, and geographic systems. That is exposure science can be applied at any level of biologic organization—ecologic, community, or individual—and, within the individual, at the level of external exposure, internal exposure, or dose. Source: Inset on exposures in space adapted from Gulliver and Briggs 2005.

physician and alchemist Paracelsus framed the widely cited toxicologic concept that "dose makes the poison" (Binswanger and Smith 2000). Ramazzini, in his 1703 treatise *Diseases of the Workers*, identified workplace exposures to single and multiple agents and the migration of contaminants into the community environment as causing disease (Ramazzini 1703). Percivall Pott first demonstrated the association between cancer and exposure to soot with his studies of scrotal cancer in chimney sweeps (Pott 1775). John Snow's study of water-use patterns and their relation to disease in London allowed him to link a source of water contamination to cholera (Snow 1885). The avoidance of potentially harmful exposures through the separation of land use between human residences and industrial facilities was proposed in the latter part of the 19th century (Howard 1898).

Use of exposure assessment in radiation health protection can be traced back to roughly 1900 after the discovery of x rays. During the 1920s, Alice Hamilton established the formal study of industrial medicine in the United States. The metrics for and applications of exposure science to radiation protection have grown in sophistication and reliability over the last century (NRC 2006; ICRP 2007; EPA 2011c). Many of the basic principles for measuring, monitoring, and modeling exposures to airborne contaminants, including the earliest use of exposure biology, come from the field of industrial hygiene. The publication of *Silent Spring* (Carson 1962) and its focus on the transfer and magnification of persistent pollutants through food webs fostered the growth of environmental toxicology and chemistry, which address chemical fate and transport through multiple media and multiple pathways.

By the middle 1980s, exposure evaluations had evolved into an established scientific discipline that moved beyond single routes, single chemicals, and single pathways toward an understanding of "total" exposure. The 1991 National Research Council report *Human Exposure Assessment for Airborne Pollutants* (NRC 1991a) laid the foundation for further development of the field by defining the core principles of exposure assessment. Between 1980 and 1985, the Total Exposure Assessment Methodology (TEAM) study was conducted to assess personal exposures of 600 residents in seven US cities to chemical exposures by one or more routes of entry into the body and to estimate the exposures and body burdens of urban populations in several cities (EPA 1987). The TEAM studies established a framework for examining total human exposure covering multiple routes of entry into the body (Wallace 1987).

By promoting the concept that it is important to "measure where the people are" (Wallace 1977), the TEAM studies revealed new source categories and control options to reduce or prevent exposures. For example, application of the concept resulted in increasing attention to exposures indoors, where people spend a substantial portion of their lives (Smith 1988a). Globally, it pointed to the importance of indoor pollution in rural areas of developing countries, where a large portion of the world's breathing is done but relatively little research or monitoring was being conducted (Smith 1988b).

Introduction

Control measures revealed by a total-exposure framework include measures to increase the time that doors are closed between a house and its garage in the United States and thereby reduce human exposure to tailpipe emissions in the home, even though this has no effect on vehicle emissions or ambient concentrations in the garage. The exposure control included a simple spring on the door to allow it to stay open a shorter time.

Two major advances that helped to establish the credibility of exposure science as a discipline were the formation of the International Society of Exposure Analysis in 1989 (now the International Society of Exposure Science) and the publication of the *Journal of Exposure Science and Environmental Epidemiology* in 1990.

A number of important milestones followed. In 1992, EPA published its *Guidelines for Exposure Assessment*, which served as a companion to its toxicology and risk-assessment guidelines. That was followed in 1993 by the initiation of the National Human Exposure Assessment Survey (NHEXAS), which evaluated human exposure to multiple chemicals on a community and regional scale (EPA 2009b). NHEXAS monitored chemicals in blood and urine; incorporated environmental sampling of air, water, soil, and dust; and conducted personal monitoring of air, food, and beverages (NRC 1991b; EPA 2009b). It brought attention to the role of the proximity of emissions as opposed to the magnitude of emissions in determining overall exposure—low-level emissions near human receptors, such as those from indoor environments, need to receive at least as much attention as outdoor stack emissions (Sexton et al. 1995). In 1997, EPA's *Exposure Factors Handbook* was published that presents data and evaluation of allometric and behavioral factors that affect exposures. It became an international resource for risk assessors who use these factors to estimate exposures for various pathways.[4]

Over the last 20 years, exposure science has evolved as a theoretical and practical science to include the development of mathematical models and other tools for examining how individuals and populations come into contact with environmental stressors of concern. For example, the discovery that airborne lead from gasoline combustion is deposited on soil, is tracked into homes, and enters children via hand-to-mouth activities greatly expanded the focus on multipathway exposure assessments and the development of exposure models that are validated through biomonitoring. Ott and others introduced time–activity models that were applied to air pollutants (Ott 1995). In the 1990s, exposure models addressed multimedia and multipathway exposures, tracking pollutants from multiple sources through air, water, soil, food, and indoor environments (McKone and Daniels 1991).

---

[4] A 2011 version has been released (EPA 2011b).

Borrowing from the concept of "dose commitment"[5] in radiation protection, researchers elaborated the concept of "exposure efficiency" in the 1980s and 1990s (for example, Smith 1993). Early in the 21st century, the term *intake fraction* was adopted to describe that concept (Bennett et al. 2002). It is defined as the amount of material crossing the body's barriers per unit emitted and thus is dimensionless. For air pollution, population intake fraction is the amount inhaled by the population divided by the amount emitted per unit activity or time. It directly connects the source and environmental-intensity boxes in Figure 1-2 with the exposure box, effectively incorporating the pathways in between without needing to specify them. A striking characteristic of intake fraction is that it varies by orders of magnitude among standard source categories—for example, in the case of air pollution, from 10^{-6} for such remote sources as power plants to 10^{-4} for urban outdoor sources, roughly 5×10^{-3} for such indoor sources as unvented stoves, and 1.0 for active smoking. Not only does "dose make the poison", therefore, but because proximity makes the dose, ultimately "place makes the poison" (NRC 2003). However, the biologically-relevant time and intensity of contact with an agent for each route of exposure needs to be considered (Lioy 1999).

OPPORTUNITIES AND CHALLENGES: THE NEW MILLENNIUM

Since 2000, a number of activities have benefited from advances in exposure science, and new challenges and opportunities have emerged. The Children's Health Act of 2000 authorized the establishment of the National Children's Study, a large-scale multiyear prospective study of children's health and exposures intended to identify and characterize environmental influences (including physical, chemical, biologic, and psychosocial) on children from birth to adulthood. The study is under way, after the completion of the Vanguard Center pilot programs and the incorporation of new tools and approaches to streamline data collection at the household level and to capitalize on existing data for constructing community exposure baselines (IOM 2008; Trasande et al. 2011).

The increasing collection and evaluation of biomarkers of exposure and effect also is providing growing opportunities for exposure science. The Centers for Disease Control and Prevention's National Health and Nutrition Examination Survey (NHANES) published the first *National Human Exposure Report* in 2001, which used a subset of its subjects to assess the US population's exposure to environmental chemicals on the basis of biomonitoring data. The reports have been updated with publications released in 2003, 2005, and 2009, and annual reports are expected. The NHANES data provide a unique and growing potential

---

[5]Dose commitment is the dose that will accumulate in an individual or population over a given period (for example, 50 years) from releases of radioactivity from a given source.

Introduction

for evaluating source–exposure and exposure–disease relationships in a national population-based representative sample. California has started its own biomonitoring program (OEHHA 2007), and other states and cities are working on biomonitoring efforts (CDC 2010). The emerging biomonitoring data sets will allow improved tracking of exposures over time, space, and across populations for an increasingly larger number of chemicals. This information will be essential for evaluating the efficacy of exposure reduction policies, and for prioritizing and assessing chemical risks.

A prime example of the benefits of improved methods of exposure assessment is their use in environmental epidemiology, in which more accurate estimates of the health effects of important stressors have been achieved by reducing exposure misclassification, for example, in air pollution (Jerrett et al. 2005) and ionizing radiation (NRC 2006). There are many opportunities for continued improvements in this arena.

The Exposome

Rapid advances in methods of sampling and analysis, genomics, systems biology, bioinformatics, and toxicology have laid the groundwork for major advances in the applications of exposure science. One such development is the concept of the "exposome", which theoretically can capture the totality of environmental exposures (including lifestyle factors, such as diet, stress, drug use, and infection) from the prenatal period on, using a combination of biomarkers, genomic technologies, and informatics (Wild 2005; Rappaport and Smith 2010). Understanding how exposures from occupation, environment, diet, lifestyle, and the like interact with unique individual characteristics—such as genetics, physiology, and epigenetic makeup resulting in disease—is the fundamental challenge implicit in the exposome. The exposome in concert with the human genome holds promise for elucidating the etiology of chronic diseases (Rappaport and Smith 2010; Wild 2012).

> **The exposome is defined as** *the record of all exposures both internal and external that people receive throughout their lifetime (Rappaport and Smith 2010).*

The concept of the exposome offers an intriguing and promising direction for exposure science that will continue to spur developments in the field, especially in biomarkers, data-sharing, and informatic approaches to large datasets. By encompassing many biomarkers and stressors at once, exposome analysis can be the source of important new hypotheses of relationships between internal markers of stress and the external environment. Within the conception of exposure science proposed here (see Figure 1-2), the committee, in Chapter 2, broadens the exposome concept to the "eco-exposome", that is the extension of exposure science from the point of contact between stressor and receptor inward into the organism and outward to the general environment, including the ecosphere.

Links to Toxicology and Risk Assessment

In recent years, the National Research Council has released two groundbreaking reports—*Toxicity Testing in the 21st Century* (NRC 2007) and *Science and Decisions: Advancing Risk Assessment* (NRC 2009)—that substantially advance conceptual and experimental approaches in the companion fields of toxicology and risk assessment. Those reports emphasize the importance of improving the assessment of early biologic markers of effects, individual susceptibility, life-stage and population vulnerability, and cumulative exposures and risks. *Toxicity Testing in the 21st Century* laid the foundation for a paradigm shift toward the use of new scientific tools to expand in vitro pathway-based toxicity testing. A key component of that report is the generation and use of population-based and individual human exposure data for interpreting test results and using toxicity biomarker data with exposure data for biomonitoring, surveillance, and epidemiologic studies. The focus of the report on systems approaches to understanding human biology, coupled with information about systems-level perturbations resulting from human–environment interactions, is critical for understanding biologically relevant exposures (Cohen Hubal 2009; Farland 2010). By emphasizing early perturbations of biologic pathways that can lead to disease, the report moved the focus of risk assessment along the exposure–disease spectrum toward exposure, especially the role of prior and current exposures in altering vulnerability of individuals and communities to additional environmental exposures. The resulting toxicology focus has essentially been on early biomarkers of effects in the population. At the same time, such concepts as the exposome have moved the focus of exposure science along the exposure–disease spectrum toward the health-effects side, especially biologic perturbations that correlate with exposure and are predictive of disease. The "meeting in the middle" carries promise for closer connections in the fields of exposure science and toxicology and for better linkages between exposure and disease (Cohen Hubal et al. 2010).

Science and Decisions: Advancing Risk Assessment, which examined ways to improve risk assessment, identified the need for better tools to address exposures in cumulative risk assessments. Its themes include the need for more and better exposure data for understanding dose–effect relationships, the need for investment in biomarkers of exposure, the importance of understanding both chemical and nonchemical stressors and their interactions, the need to use appropriate defaults to account for individual susceptibility and population vulnerability when stressor-specific data are not available, and better characterization of exposures in the context of cumulative risk assessment. The focus of *Science and Decisions: Advancing Risk Assessment* on capturing vulnerability better, on improving dose–response models, and on the observation that vulnerability arises from both prior and concurrent exposures creates important opportunities for exposure science.

Use of Exposure Science

The potential benefits of exposure science have not yet been fully realized. Among the important lags has been the slow incorporation of exposure science into policy and regulation. For example, EPA has focused on control of radon in drinking water whereas population radon exposure is actually dominated by other unregulated sources (NRC 1994; EPA 2008). Another example is the poor monitoring and control of indoor sources (for example, volatile organic compounds) even though air-pollution exposures clearly are dominated by them, as first definitively shown by the TEAM studies in the 1980s (Wallace 1991; Myers and Maynard 2005). Finally, even though occupational settings still dominate exposures to many important stressors in some populations, no effort to integrate them into population exposure-reduction strategies is under way. Political and economic barriers may help to explain those lapses, but they constitute lost opportunities to protect more people at lower cost by using exposure science (Smith 1995; Ott et al. 2007).

Integration of Human and Ecologic Exposure Science

There has been a gap been between the application of exposure science to human health and its application to ecosystem health, which is due in part to the lack of recognition of the connection between human and ecosystem health—in reality, they are inextricably linked. The connection between human health and ecosystem health is explored in the context of ecosystem services; as seen in Figure 1-4, human welfare depends on ecosystem health.

A better integration of ecologic and human exposure science is critical because ecologic conditions strongly mediate exposures and their consequences for humans and ecosystems. Not only do ecosystems contain multiple stressors that can act synergistically but organisms' environments are seldom optimal and may heighten their sensitivity to stressors. As illustrated by the examples in Box 1-2, degradation of ecosystems due to human activities increases exposure to or consequences of chemical and biologic stressors in both humans and ecosystems. Elucidating relationships between exposure and key abiotic and biotic ecologic factors is necessary if we are to understand risk.

ROADMAP

The present report builds on the concepts presented in the National Research Council reports *Toxicity Testing in the 21st Century* and *Science and Decisions: Advancing Risk Assessment* to develop a framework for bringing exposure science to a point where it fully complements toxicology and risk assessment and can be used to protect human health and the environment better. The committee also addresses a set of emerging needs, such as the need to

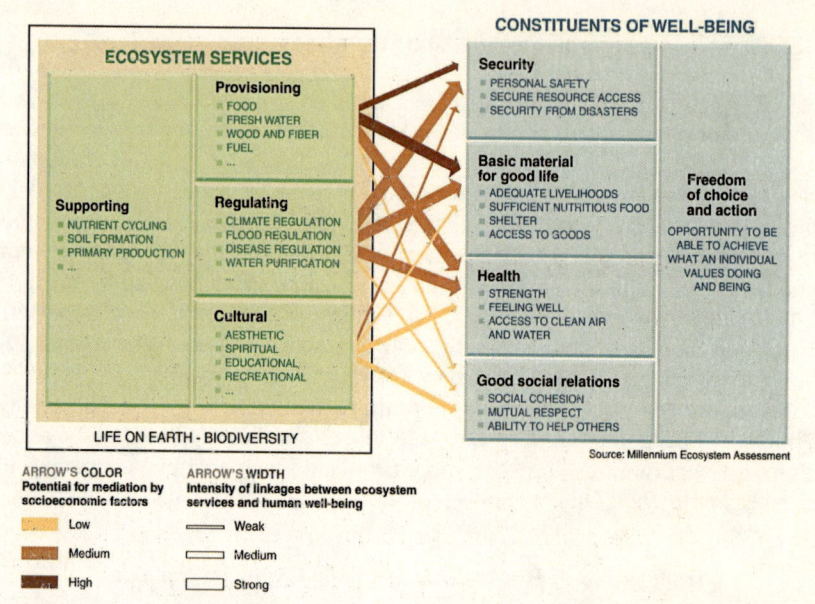

FIGURE 1-4 Connections between ecosystem services and human well-being. The framework of ecosystem services makes explicit the linkages between human and ecologic health. The strength of the linkages and the potential for mediation differ in different ecosystems and regions. Adverse exposures can indirectly affect human health and well-being by influencing a range of services provided by ecosystems. Source: Millenium Ecosystem Assessment 2005. Reprinted with permission; copyright 2005, World Resources Institute.

provide rapid assessment protocols and technologies to respond to natural and human-caused disasters and the needs for community participation and environmental justice. The report describes new technologies and opportunities to make exposure science even more effective in its traditional roles of evaluating environmental control measures, improving understanding of the link between environmental stressors and disease, and designing more cost-effective ways to reduce and prevent health risks. Finally, where possible, the committee offers ideas for integrating the applications of exposure science to human health and ecosystem health.

In Chapter 2, the committee presents a vision for exposure science. Chapter 3 describes the opportunities and challenges for applying exposure science to toxicology, epidemiology, and risk assessment and how exposure science can play a more effective role in other fields, such as environmental regulation, urban planning, ecosystem management, and disaster management. Chapter 4 addresses emerging demands for exposure-science information. Chapter 5 identifies scientific and technologic advances that are shaping the field and that support the committee's vision. Chapter 6 discusses promoting and sustaining

Introduction 33

public trust in exposure science including the management of personal and environmental exposure data. Chapter 7 describes the path forward for exposure science in the 21st century.[6]

BOX 1-2 Illustrations Demonstrating How the Degradation of Ecosystems Due to Human Activities Increases Exposures to Chemical and Biologic Stressors

Rising temperatures. Whether caused by shifts in climate or land uses (for example, deforestation, reduced vegetative cover, and urban heat islands), changes in temperature can directly prompt health-threatening exposures (for example, extreme heat events) or indirectly influence exposure to other substances. In aquatic ecosystems, degraded riparian zones, loss of forest cover, runoff from impervious surfaces, and discharges from industry can lead to rising water temperatures and increased toxicity. Above-normal temperatures compromise function and integrity of aquatic ecosystems. In addition, high temperatures can increase sensitivity of aquatic animals to heavy metals, including cadmium (Lannig et al. 2006; Cherkasov et al. 2006, 2007), mercury (Slotsbo et al. 2009), copper (Gupta et al.1981; Boeckman and Bidwell 2006; Khan et al. 2006), and lead (Khan et al. 2006). High temperatures also may amplify effects of pesticides—such as diazinon (Osterauer and Köhler 2008), terbufos, and trichlorfon (Brecken-Folse et al. 1994; Howe et al. 1994)—on fish.

Anthropogenic nutrient enrichment. Agricultural runoff and untreated sewage effluent are two important causes of eutrophication, in which aquatic ecosystems accumulate high concentrations of nutrients (for example, phosphates and nitrates) that promote plant growth. Algal growth can become excessive and sometimes lead to harmful algal blooms (Paerl 1997; Cloern 2001; Anderson et al. 2002; Kemp et al. 2005) and anoxic (low-oxygen) conditions that directly kill organisms and that can increase sensitivity to chemical stressors. For example, low dissolved oxygen prompted higher mortality in daphnids exposed to carbendazim (Ferreira et al. 2008), in crabs exposed to copper (Depledge 1987), and in fish exposed to alkylphenols (Gupta et al. 1983).

Reduced access to water. Human-associated changes in hydrologic regimes—including construction of dams and levees, depletion of groundwater supplies, drainage of wetlands, and removal of vegetation— profoundly affect water availability for humans and ecologic communities alike. Aside from the direct effects on ecosystem goods and services related to water, these an-

(Continued)

---

[6]Given its terms of reference, the committee addresses primarily exposure-science issues related to the United States and other developed countries. In addition, the committee does not focus on uses of exposure principles in regulation and policy.

BOX 1-2 Continued

thropogenic stressors can promote dehydration, which can increase concentration of toxicants and thereby increase risk of damage. Chemicals also can reduce drought tolerance of organisms by interfering with physiologic adaptations, as has been demonstrated in earthworms exposed to copper (Holmstrup 1997) and in springtails exposed to polyclic aromatic hydrocarbons (Sjursen et al. 2001), lidane (Demon and Eijsackers 1985), and surfactants (Holmstrup 1997; Skovlund et al. 2006). Diminishing access to safe water can increase risk of some diseases as wildlife, livestock, and humans are brought into closer contact.

Invasive species. Biotic invasion is one of the top drivers of biodiversity loss and species endangerment. Invasive species can alter species interactions and disrupt ecologic processes in ways that elicit serious ecologic, economic, and health consequences. Even seemingly benign species can provoke unexpected exposures. For example, a recent experiment suggested that Amur honeysuckle (*Lonicera maackii*), a widespread invasive shrub in North America, increases human risk of exposure to ehrlichiosis, an emerging infectious disease transmitted by ticks (Allan et al. 2010). The high risk would result from a preference of a key tick and pathogen reservoir, white-tailed deer (*Odocoileus virginianus*), for areas of dense honeysuckle. In aquatic systems, the invasive round goby (*Neogobius melanostomus*) is thought to facilitate mobilization of contaminants in food webs and to increase exposure to humans because its persistence in contaminated environments draws predatory fish, which also are popular game species, into polluted habitats (Marentette et al. 2010).

Shifts in species composition. Because species differ in bioaccumulation kinetics, changes in the structure of animal communities can influence bioaccumulation and human exposure. Indeed, mercury accumulation rates differ among bivalve species according to feeding strategies and assimilation efficiencies (Cardoso et al. 2009). At the terrestrial–aquatic interface, spiders had more of the highly bioavailable methylmercury than other invertebrates (such as lepidopterans and orthopterans) and, therefore were thought to be responsible for transporting aquatic mercury into terrestrial food webs (Cristol et al. 2008). The presence of particular species can provide buffers to exposure in some cases; for example, some algal blooms are known to reduce uptake of methylmercury into freshwater food webs (Pickhardt et al. 2002).

Habitat degradation. Habitat degradation can facilitate transport of contaminants into aquatic systems, transmission of diseases by promoting high densities of vectors, and increases in the sensitivity of animals to exposures. Changes in food availability in degraded habitats also can affect nutritional status in ways that can mediate toxicity (Holmstrup et al. 2010). Consequences of habitat degradation can be surprising. For example, erosion of

(Continued)

BOX 1-2 Continued

European saltmarshes dominated by cord grasses has resulted in massive cadmium release from sediments in areas of cord-grass dieback (Hubner et al. 2010); this shows how habitat degradation or sea-level rise, for example, can increase exposure to heavy metals.

Emerging contaminants. A number of new compounds with novel chemistries are rapidly expanding in commerce and are now appearing, or are expected to occur, as widespread environmental contaminants. Important examples include brominated and chlorinated organic compounds used as flame retardants and a large and growing array of diverse organic and metal-based nanomaterials (Lorber 2008; Stapleton et al. 2006, 2009; Wiesner et al. 2009). Little is known about human health or ecologic effects of most of these materials. Similarly, little is known about fates and exposures; it is difficult to measure most of them at environmentally relevant concentrations in media and in organisms. The situation is particularly problematic with nanomaterials. It is likely that as their abundance increases in the environment, they will contribute to the sum of multiple stressors to which humans and ecosystems are exposed. And their fates and effects will be influenced by the environmental variables described above. It is timely for exposure science to be rapidly developing concomitantly with these new chemistries.

Greenhouse-gas emissions. Increased concentrations of atmospheric carbon dioxide stimulate growth of poison ivy (*Toxicodendron radicans*) and causes plants to produce a more allergenic form of urushiol (Mohan et al., 2006). Anthropogenic changes in the atmosphere thus are expected to lead to more abundant and more toxic poison ivy.

REFERENCES

Allan, B.F., H.P. Dutra, L.S. Goessling, K. Barnett, J.M. Chase, R.J. Marquis, G. Pang, G.A. Storch, R.E. Thach, and J.L. Orrock. 2010. Invasive honeysuckle eradication reduces tick-borne disease risk by altering host dynamics. Proc. Natl. Acad. Sci. U. S. A. 107(43):18523-18527.

Anderson, D.M., P.M. Glibert, and J.M. Burkholder. 2002. Harmful algal blooms and eutrophication: Nutrient sources, composition, and consequences. Estuaries 25(4B):704-726.

Bennett, D.H., T.E. McKone, J.S. Evans, W.W. Nazaroff, M.D. Margni, O. Jolliet, and K.R. Smith. 2002. Defining intake fraction. Environ. Sci. Technol. 36(9):206A-211A.

Binswanger, H.C., and K.R. Smith. 2000. Paracelsus and Goethe: Founding fathers of environmental health. Bull. World Health Organ. 78(9):1162-1164.

Boeckman, C.J., and J.R. Bidwell. 2006. The effects of temperature, suspended solids, and organic carbon on copper toxicity to two aquatic invertebrates. Water Air Soil Pollut. 171(1-4):185-202.

Brecken-Folse, J.A., F.L. Mayer, L.E. Pedigo, and L.L. Marking. 1994. Acute toxicity of 4-nitrophenol, 2,4-dinitrophenol, terbufos and trichlorofon to grass shrimp (*Palaemonetes Spp.*) and sheepshead minnows (*Cyprinodon variegatus*) as affected by salinity and temperature. Environ. Toxicol. Chem. 13(1):67-77.

Cardoso, P.G., A.I. Lillebo, E. Pereira, A.C. Duarte, and M.A. Pardal. 2009. Different mercurybioaccumulation kinetics by two macrobenthic species: The bivalve *Scrobicularia plana* and the polychaete *Hediste diversicolor*. Mar. Environ. Res. 68(1):12-18.

Carson, R. 1962. Silent Spring. Cambridge, MA: Riverside Press.

CDC (Centers for Disease Control and Prevention). 2010. National Biomonitoring Program, State Grant Activities [online]. Available: http://www.cdc.gov/biomonitoring/state_grants.html [accessed Dec. 6, 2011].

Cherkasov, A.S., P.K. Biswas, D.M. Ridings, A.H. Ringwood, and I.M. Sokolova. 2006. Effects of acclimation temperature and cadmium exposure on cellular energy budgets in the marine mollusk *Crassostrea virginica*: Linking cellular and mitochondrial responses. J. Exp. Biol. 209(7):1274-1284.

Cherkasov, A.S., S. Grewal, and I.M. Sokolova. 2007. Combined effects of temperature and cadmium exposure on haemocyte apoptosis and cadmium accumulation in the eastern oyster *Crassostrea virginica* (Gmelin). J. Therm. Biol. 32(3):162-170.

Cloern, J.E. 2001. Our evolving conceptual model of the coastal eutrophication problem. Mar. Ecol. Progr. Ser. 210:223-253.

Cohen Hubal, E.A. 2009. Biologically relevant exposure science for 21st century toxicity testing. Toxicol. Sci.111(2):226-232.

Cohen Hubal, E.A., A. Richard, L. Aylward, S. Edwards, J. Gallagher, M.R. Goldsmith, S. Isukapalli, R. Tornero-Velez, E. Weber, and R. Kavlock. 2010. Advancing exposure characterization for chemical evaluation and risk assessment. J. Toxicol. Environ. Health B Crit. Rev. 13(2-4):299-313.

Cristol, D.A., R.L. Brasso, A.M. Condon, R.E. Fovargue, S.L. Friedman, K.K. Hallinger, A.P. Monroe, and A.E. White. 2008. The movement of aquatic mercury through terrestrial food webs. Science 320(5874):335-335.

Demon, A., and H.Eijsackers. 1985. The effects of lindane and azinphosmethyl on survival-time of soil animals, under extreme or fluctuating temperature and moisture conditions. Z. Angew. Entomol. 100(1-5):504-510.

Depledge, M.H. 1987. Enhanced copper toxicity resulting from environmental-stress factor synergies. Comp. Biochem. Physiol. C-Pharmacol. 87(1):15-19.

Eisenberg, J.N., X. Lei, A.H. Hubbard, M.A. Brookhart, and J.M. Colford, Jr. 2005. The role of disease transmission and conferred immunity in outbreaks: Analysis of the 1993 *Cryptosporidium* outbreak in Milwaukee, Wisconsin. Am. J. Epidemiol. 161(1):62-72.

EPA (U.S. Environmental Protection Agency). 1987. Project Summary: The Total Exposure Assessment Methodology (TEAM) Study. EPA 600/S6-87/002. Office of Acid Deposition, Environmental Monitoring and Quality Assurance, Office of Research and Development, U.S. Environmental Protection Agency Washington, DC. September 1987.

EPA (U.S. Environmental Protection Agency). 2003. Framework for Cumulative Risk Assessment. EPA/630/P-02/001. Risk Assessment Forum, U.S. Environmental Protection Agency, Washington, DC [online]. Available: http://www.epa.gov/raf/publiccations/pdfs/frmwrk_cum_risk_assmnt.pdf [accessed Dec. 29, 2011].

EPA (U.S. Environmental Protection Agency). 2008. Care for Your Air: A Guide to Indoor Air Quality. EPA 402/F-08/008. U.S. Environmental Protection Agency, September 2008 [online]. Available: http://www.epa.gov/iaq/pdfs/careforyourair.pdf [accessed April 9, 2012].

EPA (U.S. Environmental Protection Agency). 2009a. A Conceptual Framework for U.S. EPA's National Exposure Research Laboratory. EPA/600/R-09/003. National Exposure Research Laboratory, Office of Research and Development, U.S. Environmental Protection Agency [online]. Available: http://www.epa.gov/nerl/documents/nerl_exposure_framework.pdf [accessed Dec. 29, 2011].

EPA (U.S. Environmental Protection Agency). 2009b. Human Exposure Measurements: National Human Exposure Assessment Survey (NHEXAS). U.S. Environmental Protection Agency [online]. Available: http://www.epa.gov/heasd/edrb/nhexas.html [accessed Dec. 29, 2011].

EPA (U.S. Environmental Protection Agency). 2011a. EPA Risk Assessment Basic Information: What is Risk? What is a Stressor? U.S. Environmental Protection Agency [online]. Available: http://www.epa.gov/risk_assessment/basicinformation.htm#risk [accessed Dec. 29, 2011].

EPA (U.S. Environmental Protection Agency). 2011b. Exposure Factors Handbook: 2011 Edition. EPA/600/R-090/052F. National Center for Environmental Assessment, Office of Research and Development, U.S. Environmental Protection Agency, Washington, DC [online]. Available: http://www.epa.gov/ncea/efh/pdfs/efh-complete.pdf [accessed Oct. 27, 2011].

EPA (U.S. Environmental Protection Agency). 2011c. EPA Radiogenic Cancer Risk Models and Projections for the U.S. Population. EPA 402-R-11-001. Office of Radiation and Indoor Air, U.S. Environmental Protection Agency, Washington, DC [online]. Available: http://www.epa.gov/rpdweb00/docs/bluebook/402-r-11-01.pdf [accessed Nov. 2, 2011].

EPA SAB (U.S. Environmental Protection Agency, Science Advisory Board). 1992. Respiratory Health Effects of Passive Smoking: Lung Cancer and Other Disorders, SAB Review Draft. EPA /600/6-90/0068. Science Advisory Board, U.S. Environmental Protection Agency, Washington, DC. May 1992.

Farland, W.H. 2010. The promise of exposure science and assessment. J. Expo. Sci. Environ. Epidemiol. 20(3):225.

Ferreira, A.L.G., S. Loureiro, and A.M. Soares. 2008. Toxicity prediction of binary combinations of cadmium, carbendazim and low dissolved oxygen on *Daphnia magna*. Aquat. Toxicol. 89(1):28-39.

GAO (General Accounting Office). 2005. Chemical regulation: Options Exist to Improve EPA's Ability to Assess Health Risks and Manage its Chemical Review Program. GAO-05-458. Washington, DC: U.S. General Accounting Office [online]. Available: http://www.gao.gov/new.items/d05458.pdf [accessed Dec. 29, 2011].

Graham, J.A., ed. 2011. Exposure Science Digests: Demonstrating How Exposure Science Protects Us From Chemical, Physical, and Biological Agents. Journal of Exposure Science and Environmental Epidemiology [online]. Available: http://www.nature.com/jes/pdf/JESSE_ESD_booklet.pdf [accessed Dec. 29, 2011].

Gulliver, J., and D.J. Briggs. 2005. Time-space modelling of journey-time exposure to traffic-related air pollution using GIS. Environ. Res. 97(1):10-25.

Gupta, P.K., B.S. Khangarot, and V.S. Durve. 1981. The temperature-dependence of the acute toxicity of copper to a fresh-water pond snail, *Viviparus bengalensis* L. Hydrobiologia 83(3):461-464.

Gupta, S., R.C. Dalela, and P.K. Saxena. 1983. Influence of dissolved-oxygen levels on acute toxicity of phenolic-compounds to fresh-water teleost, *Notopterus notopterus* (Pallas). Water Air Soil Pollut. 19(3):223-228.

Holmstrup, M. 1997. Drought tolerance in *Folsomia candida Willem* (Collemhola) after exposure to sublethal concentrations of three soil-polluting chemicals. Pedobiologia 41(4):361-368.

Holmstrup, M., A. Bindesbol, G.J. Oostingh, A. Duschl, V. Scheil, H. Köhler, S. Loureiro, A.M. Soares, A.L. Ferreira, C. Kienle, A. Gerhardt, R. Laskowski, P.E. Kramarz, M. Bayley, C. Svendsen, and D.J. Spurgeon. 2010. Interactions between effects of environmental chemicals and natural stressors: A review. Sci. Total Environ. 408(18):3746-3762.

Howard, E. 1898. Tomorrow; A Peaceful Path to Real Reform. London: S. Sonnenschein.

Howe, G.E., L.L. Marking, T.D. Bills, J.J. Rach, and F.L. Mayer, Jr. 1994. Effects of water temperature and pH on toxicity of terbufos, trichlorfon, 4-nitrophenol and 2,4-dinitrophenol to the amphipod *Gammarus pseudolimnaeus* and rainbow trout (*Oncorhynchus mykiss*). Environ. Toxicol. Chem. 13(1):51-66.

ICRP (International Commission on Radiological Protection). 2007. The 2007 Recommendations of the International Commission on Radiological Protection. Annals of the ICRP 103 [online]. Available: http://www.icrp.org/docs/ICRP_Publication_103-Annals_of_the_ICRP_37%282-4%29-Free_extract.pdf [accessed Nov. 2, 2011].

Hűbner, R., R.J.H. Herbert, and K.B. Astin. 2010. Cadmium release caused by the dieback of the saltmarsh cord grass *Spartina anglica* in Poole Harbour (UK). Estuar. Coast. Shelf S. 87(4):553-560.

IOM (Institute of Medicine). 2000. Protecting Those Who Serve: Strategies to Protect the Health of Deployed U.S. Forces. Washington, DC: National Academy Press.

IOM (Institute of Medicine). 2008. The National Children's Study Research Plan: A Review. Washington, DC: National Academies Press.

Jerrett, M., R.T. Burnett, R. Ma, C.A. Pope, D. Krewski, K.B. Newbold, G. Thurston, Y. Shi, N. Finkelstein, E.E. Calle, and M.J. Thun. 2005. Spatial analysis of air pollution and mortality in Los Angeles. Epidemiology 16(6):727-736.

Jones, R.L., D.M. Homa, P.A. Meyer, D.J. Brody, K.L. Caldwell, J.L. Pirkle, and M.J. Brown. 2009. Trends in blood lead levels and blood lead testing among US children aged 1 to 5 years, 1988-2004. Pediatrics 123(3):e376-e385.

Kemp, W.M., W.R. Boynton, J.E. Adolf, D.F. Boesch, W.C. Boicourt, G. Brush, J.C. Cornwell, , T.R. Fisher, P.M. Glibert, J.D. Hagy, L.W. Harding, E.D. Houde, D.G. Kimmel, W.D. Miller, R.I.E. Newell, M.R. Roman, E.M. Smith, and J.C. Stevenson. 2005. Eutrophication of Chesapeake Bay: Historical trends and ecological interactions. Mar. Ecol. Progr. Ser 303:1-29.

Khan, M.A., S.A. Ahmed, B. Catalin, A. Khodadoust, O. Ajayi, and M. Vaughn. 2006. Effect of temperature on heavy metal toxicity to juvenile crayfish, *Orconectes immunis* (Hagen). Environ. Toxicol. 21(5):513-520.

Lannig, G., A.S. Cherkasov, and I.M. Sokolova. 2006. Temperature-dependent effects of cadmium on mitochondrial and whole-organism bioenergetics of oysters (*Crassostrea virginica*). Mar. Environ. Res. 62(suppl.):S79-S82.

Lioy, P.J. 1990. Assessing total human exposure to contaminants: A multidisciplinary approach. Environ. Sci. Technol. 24(7):938-945.

Lioy, P.J. 1999. The 1998 ISEA Wesolowski Award lecture. Exposure analysis: reflections on its growth and aspirations for its future. J. Expo. Anal. Environ. Epidemiol. 9(4):273-281.

Lorber, M. 2008. Exposure of Americans to polybrominated diphenyl ethers. J. Exp. Sci. Environ. Epidemiol. 18(1):2-19.

Marentette, J.R., K.L. Gooderham, M.E. McMaster, T. Ng, J.L. Parrott, J.Y. Wilson, C.M. Wood, and S. Balshine. 2010. Signatures of contamination in invasive round gobies (*Neogobius melanostomus*): A double strike for ecosystem health? Ecotoxicol. Environ. Saf. 73(7):1755-1764.

McKone, T.E., and J.I. Daniels. 1991. Estimating human exposure through multiple pathways from air, water, and soil. Regul. Toxicol. Pharmacol. 13(1):36-61.

Millennium Ecosystem Assessment. 2005. Ecosystems and Human Well-being: Synthesis. Island Press, Washington, DC [online]. Available: http://www.maweb.org/documents/document.356.aspx.pdf [accessed Feb. 8, 2011].

Mohan, J.E., L.H. Ziska, W.H. Schlesinger, R.B. Thomas, R.C. Sicher, K. George, and J.S. Clark. 2006. Biomass and toxicity responses of poison ivy (*Toxicodendron radicans*) to elevated atmospheric CO_2. Proc. Natl. Acad. Sci. U. S. A. 103(24):9086-9089.

Muntner, P., A. Menke, K.B. DeSalvo, F.A. Rabito, and V. Batuman. 2005. Continued decline in blood lead levels among adults in the United States: The National Health and Nutrition Examination Surveys. Arch. Intern. Med. 165(18):2155-2161.

Myers, I., and R.L. Maynard. 2005. Polluted air—outdoors and indoors. Occup. Med. 55(6):432-438.

Myers, M.S., L.L. Johnson, and T.K. Collier. 2003. Establishing the causal relationship between polycyclic aromatic hydrocarbon (PAH) exposure and hepatic neoplasms and neoplasia-related liver lesions in English sole (*Pleuronectes vetulus*). Hum. Ecol. Risk Assess. 9(1):67-94.

NRC (National Research Council). 1991a. Human Exposure Assessment for Airborne Pollutants: Advances and Opportunities. Washington, DC: National Academy Press.

NRC (National Research Council). 1991b. Monitoring Human Tissues for Toxic Substances. Washington, DC: National Academy Press.

NRC (National Research Council). 1994. Health Effects of Radon: Time for Reassessment? Washington, DC: The National Academy Press.

NRC (National Research Council). 1998. Research Priorities for Airborne Particulate Matter: I. Immediate Priorities and a Long-Range Research Portfolio. Washington, DC: The National Academy Press.

NRC (National Research Council). 1999. Our Common Journey: A Transition Toward Sustainability. Washington, DC: National Academy Press.

NRC (National Research Council). 2003. Managing Carbon Monoxide Pollution in Meteorological and Topographical Problem Areas. Washington, DC: National Academies Press.

NRC (National Research Council). 2006. Health Risks from Exposure to Low Levels of Ionizing Radiation: BEIR VII Phase 2. Washington, DC: National Academies Press.

NRC (National Research Council). 2007. Toxicity Testing in the 21st Century: A Vision and A Strategy. Washington, DC: The National Academies Press.

NRC (National Research Council). 2009. Science and Decisions: Advancing Risk Assessment. Washington, DC: National Academies Press.

OEHHA (Office of Environmental Health Hazard Assessment). 2007. California Environmental Contaminant Biomonitoring Program. Office of Environmental Health Hazard Assessment, California Environmental Protection Agency [online]. Available: http://oehha.ca.gov/multimedia/biomon/about.html [accessed Dec. 6, 2011].

Osterauer, R., and H. Köhler. 2008. Temperature-dependent effects of the pesticides thiacloprid and diazinon on the embryonic development of zebrafish (*Danio rerio*). Aquat. Toxicol. 86(4):485-494.

Ott, W.R. 1995. Human exposure assessment: The birth of a new science. J. Expo. Sci. Environ. Epidemiol. 5(4):449-472.

Ott, W., A.C. Steinemann, and L.A. Wallace, eds. 2007. Exposure Analysis. Boca Raton: CRC Press.

Paerl, H.W. 1997. Coastal eutrophication and harmful algal blooms: Importance of atmospheric deposition and groundwater as "new" nitrogen and other nutrient sources. Limnol. Oceanogr. 42(5):1154-1165.

Pastor, M., R. Morello-Frosch, and J.L. Sadd. 2005. The air is always cleaner on the other side: Race, space, and ambient air toxics exposures in California. J. Urban Aff. 27(2):127-148.

Pickhardt, P.C., C.L. Folt, C.Y. Chen, B. Klaue, and J.D. Blum. 2002. Algal blooms reduce the uptake of toxic methylmercury in freshwater food webs. Proc. Natl. Acad. Sci. U. S. A. 99(7):4419-4423.

Pott, P. 1775. Cancer Scroti. Pp. 63-68 in Chirurgical Observations Relative to the Cataract, the Polypus of the Nose, the Cancer of the Scrotum, the Different Kinds of Ruptures and the Mortification of the Toes and Feet. London: T.J. Carnegy.

Prüss-Üstün, A., and C. Corvalán. 2006. Preventing Disease Through Healthy Environments: Towards an Estimate of the Environmental Burden of Disease. Geneva: World Health Organization [online]. Available: http://www.who.int/quantifying_ ehimpacts/publications/preventingdisease.pdf [accessed Dec. 29, 2011].

Rappaport, S.M., and M.T. Smith. 2010. Environment and disease risks. Science 30(6003):460-461.

Ramazzini, B. 1703. De morbis artificum diatribe. Ultrajecti: Apud Guilielmum van de Water.

Sexton, K., M.A. Callahan, and E.F. Bryan. 1995. Estimating exposure and dose to characterize health risks: The role of human tissue monitoring in exposure assessment. Environ. Health Perspect. 103(suppl. 3):13-29.

Sjursen, H., L.E. Sverdrup, and P.H. Krogh. 2001. Effects of polycyclic aromatic compounds on the drought tolerance of *Folsomia fimetaria* (collembola, isotomidae). Environ. Toxicol. Chem. 20(12):2899-2902.

Skovlund, G., C. Damgaard, M. Bayley, and M. Holmstrup. 2006. Does lipophilicity of toxic compounds determine effects on drought tolerance of the soil collembolan *Folsomia candida*? Environ. Pollut. 144(3):808-815.

Slotsbo, S., L.H. Heckmann, C. Damgaard, D. Roelofs, T. de Boer, and M. Holmstrup. 2009. Exposure to mercury reduces heat tolerance and heat hardening ability of the springtail *Folsomia candida*. Comp. Biochem. Physiol. C Toxicol. Pharmacol. 150(1):118-123.

Smith, K.R. 1988a. Total exposure assessment: Part 1. Implications for the U.S. Environment 30(8):10-15; 33-38.

Smith, K.R. 1988b. Total exposure assessment: Part 2. Implications for developing countries. Environment 30(10):16-20; 28-35.

Smith, K.R. 1993. Fuel combustion, air pollution exposure, and health: The situation in developing countries. Annu. Rev. Energ. Environ. 18:529-566.

Smith, K.R. 1995. The Potential of Human Exposure Assessment for Air Pollution Regulation. Human Exposure Assessment Series WHO/EHG/95.09. Geneva: World Health Organization.

Smith, K.R., C.F. Corvalán, and T. Kjellström. 1999. How much global ill health is attributable to environmental factors? Epidemiology 10(5):573-584.

Snow, J. 1855. On the Mode of Communication of Cholera, 2nd Ed. London: Churchill.

Stapleton, H.M, N.G. Dodder, J.R. Kucklick, C.M. Reddy, M.M. Schantz, P.R. Becker, F. Gulland, B.J. Porter, and S.A. Wise. 2006. Determination of HBCD, PBDEs and MeO-BDEs in California sea lions (*Zalophus californianus*) stranded between 1993 and 2003. Mar. Pollut. Bull. 52(5):522-531.

Stapleton, H.M., S. Klosterhaus, S. Eagle, J. Fuh, J.D. Meeker, A. Blum, and T.F. Webster. 2009. Detection of organophosphate flame retardants in furniture foam and U.S. house dust. Environ. Sci. Technol. 43(19):7490-7495.

Trasande, L., H.F. Andrews, C. Goranson, W. Li, E.C. Barrow, S.B. VenderBeek, B. McCrary, S.B. Allen, K.D. Gallagher, A. Rundle, J. Quinn, and B. Brenner. 2011. Early experiences and predictors of recruitment success for the National Children's Study. Pediatrics 127(2):261-268.

Wallace, L.A. 1977. Personal air quality monitors. Pp. 61-91 in Analytical Studies for the U.S. Environmental Protection Agency, Vol. IVa. Environmental Monitoring Supplement. Washington, DC: National Academy of Sciences.

Wallace, L.A. 1987. The Total Exposure Assessment Methodology (TEAM) Study: Summary and Analysis, Vol. 1. EPA/600/6-87/002a. Office of Research and Development, U.S. Environmental Protection Agency, Washington, DC [online]. Available: http://exposurescience.org/pub/reports/TEAM_Study_book_1987.pdf [accessed Nov. 4, 2011].

Wallace, L.A. 1991. Comparison of risks from outdoor and indoor exposure to toxic chemicals. Environ. Health Perspect. 95:7-13.

Wasserstein, A. 1982. Galen's Commentary on the Hippocratic Treatise Airs, Waters, Places in the Hebrew Translation of Solomon ha-Me'ati. Proceedings of the Israel Academy of Sciences and Humanities 6(3). Jerusalem: Israel Academy of Sciences and Humanities.

WHO (World Health Organization). 2004. World Health Report-2004: Changing History. Geneva: World Health Organization [online]. Available: http://www.who.int/whr/2004/en/report04_en.pdf [accessed Dec. 29, 2011].

Wiesner, M.R., G.V. Lowry, K.L. Jones, M.F. Hochella, Jr., R.T. Di Giulio, E. Casman, and E.S. Bernhardt. 2009. Decreasing uncertainties in assessing environmental exposure, risk and ecological implications of nanomaterials. Environ. Sci. Technol. 43(17):5458-6462.

Wild, C.P. 2005. Complementing the genome with an "exposome": The outstanding challenge of environmental exposure measurement in molecular epidemiology. Cancer Epidemiol. Biomarkers Prev. 14(8):1847-1850.

Wild, C.P. 2012. The exposome: From concept to utility. Int. J. Epidemiol. 41(1):24-32.

2

A Vision for Exposure Science in the 21st Century

Understanding the contact between a stressor and a receptor is at the heart of exposure science—and the starting point for the committee's expanded vision for exposure science in the 21st century. Federal, state, and municipal agency use of the concept of exposure to control environmental risks has contributed to major improvements in environmental protection in the United States and elsewhere. For example, the concept of exposure was instrumental in efforts to control secondhand tobacco smoke, a major source of exposure, but a minor contributor to emissions or ambient air pollution where measurement and regulation had focused previously.

New challenges and new scientific advances, documented in Chapters 4 and 5, impel us to an expanded vision of exposure science. Understanding aggregate or cumulative exposures in their full environmental context will require new approaches to exposure assessment—moving both inward and outward from the core point of contact.

We modify the exposome concept described in Chapter 1 in this broader vision of exposure science:

- **Vision:** exposure science extends from the point of contact between stressor and receptor inward into the organism and outward to the general environment, including the ecosphere.

We suggest the term "eco-exposome" to encapsulate the concept of this expanded vision.

In light of the new concept, we foresee, among other developments, the evolution of a universal exposure-tracking framework that allows the creation of an exposure narrative and the prediction of virtually all biologically relevant human and ecologic exposures with sensitivity, specificity, and wide coverage.

To retain its value, the principal goal of exposure science must continue to be the prevention and mitigation of adverse exposures to protect human and ecosystem health with a focus on the routes by which environmental stressors reach humans and ecosystems.

Given that vision, exposure science can maintain and augment its relevance to the everyday lives of citizens as they seek measures to protect their health and the health of the ecosystems on which they depend. Exposure science also will continue to contribute to understanding of exposures in numerous settings—outdoors, indoors, and in the occupational environment—and of how those exposures internalize in an organism.

Implicit in the eco-exposome concept is the recognition that humans live in and are part of an ecosphere and that human exposures are intimately linked to exposures flowing through ecosystems. Efforts to understand ecosystem exposures will improve our knowledge of how exposures affect human populations and individuals. Narrating the flow and pulse of exposures through the ecosphere, of which humans are part, also promotes a more thorough investigation of the potential sources of exposure and how these sources can be controlled to protect public and ecosystem health. Because the life courses of humans and ecosystems are dynamic, the eco-exposome will have to evolve scientifically to define what constitutes a biologically relevant exposure.

The committee's vision is premised on scientific developments of the last decade. Advances in local sensor systems, remote sensing, analytic methods, molecular technologies, computational modeling systems, and bioinformatics have provided opportunities to develop systems approaches that can be integrated into exposure science. There is now an unprecedented opportunity to consider exposures from source to dose, on multiple levels of integration within the ecosphere (including time, space, and biologic scales), to multiple stressors, and scaled from molecular systems to individuals, populations, and ecosystems.

Many of the scientific innovations have been in fields outside traditional exposure science, and achieving the vision will require higher levels of transdisciplinary and interagency cooperation than have occurred in the field of exposure science in the past. In addition, collaborative approaches will be needed to engage communities and stakeholders from problem formulation through data collection to development of responsive solutions and to improve communication among and participation by stakeholders. Such engagement strategies in field studies can lead to more comprehensive application of exposure-science tools to health and environmental protection, including issues of environmental justice.

In this report, the committee presents a roadmap of how technologic innovations and strategic collaborations can advance exposure science in the 21st century. The committee believes that exposure science needs to deliver knowledge that is effective, timely, and relevant to current and future environmental-health challenges. To do so, exposure science needs to continue to build capacity to

- **Assess and mitigate** exposures quickly in the face of emerging environmental-health threats and natural and human-caused disasters.
- **Predict and anticipate** human and ecologic exposures related to existing and emerging threats.
- **Customize solutions** that are scaled to identified problems.
- **Engage stakeholders** associated with the development, review, and use of exposure-science information, including regulatory and health agencies and groups that might be disproportionately affected by exposures.

The committee recognizes the complex interdependence of human and ecologic systems and adopts an integrated definition of environmental health. In the context of the vision statement, exposure science addresses chemical, physical, and biologic stressors and associated behavioral and societal factors that affect human and ecologic health, including protection of vulnerable populations and susceptible individuals.

The following sections elaborate on core elements of the vision. Later chapters provide specific components of the roadmap for realizing the vision of exposure science in the 21st century.

Assess and Mitigate

Assessing and mitigating exposures effectively require techniques for rapid measurement of single and multiple stressors on diverse geographic, temporal, and biologic scales and an enhanced infrastructure for rapid deployment of resources to address imminent threats, such as the Deepwater Horizon oil spill, Hurricane Katrina, and the tragedy of 9/11. In those three cases, there was a need to evaluate the status of the environment and the exposure of populations via a variety of pathways (such as air, water, soil, and food) while anticipating potential health effects. In the immediate aftermath of Hurricane Katrina, for example, there was an urgent need to evaluate drinking-water safety; a wide array of potential microbiologic and chemical contaminants in sediment, soil, and fish; and air-quality threats posed by mold, endotoxins, and other contaminants in indoor and outdoor environments. It is important that such assessments be handled quickly and effectively to inform and protect first responders, cleanup workers, and affected populations and to respond to stakeholder concerns about the potential for short-term and long-term health effects.

The use of more portable instruments and new techniques in biologic and environmental monitoring will enable faster identification of chemical, biologic, or physical stressors that are affecting humans or ecosystems. Testing of stressors of potential concern in targeted studies would allow rapid responses and deployment of exposure-mitigation measures.

Predict and Anticipate

Enhancing our predictive capabilities through the development of models or modeling systems will enable us to anticipate exposures and characterize exposures that had not been previously considered. For example, modeling will improve our ability to reconstruct external exposures on the basis of the increasing number of internal markers of exposures that are being collected. In addition, exposure models and controlled simulation studies will enable sustainable innovation in developing benign nanomaterials and less toxic chemical alternatives. Predictive tools will also allow us to develop exposure information on thousands of chemicals that are now in widespread use and will enable informed safety assessments of existing and new applications for these chemicals. Finally, predictive tools will allow us to forecast, prevent, and mitigate the potential effects of major societal problems, such as climate change, security threats, and urbanization. Innovative and expedient exposure-assessment approaches that strategically use diverse information such as structural properties of chemicals, nontargeted environmental surveillance, biomonitoring, and modeling and related data-integration tools are needed for the identification and quantification of relevant exposures that may pose a threat to ecosystems or human health.

Using such tools, especially in parallel with pathway-based toxicity screening, in the evaluation of new and emerging environmental stressors can help to ensure that substances in the marketplace are safer. The tools are likely to require broader access to data that now are commonly proprietary, including manufacturing, import, sales, and use data and chemical properties. New data-generation requirements also may be needed; for example, systematic toxicity screening and screening-level exposure assessments would provide a more robust basis for modeling. Exposure-based predictive screening could identify and predict introduction of chemicals that pose potentially serious environmental or health concerns. For example, thoughtful application of predictive tools would have prevented hasty promotion of methyl tertiary butyl ether as a replacement for lead in gasoline, which resulted in widespread contamination of groundwater, or universal application of polybrominated diphenyl ethers as flame retardants, which are now ubiquitous in trace amounts in human breast milk (Goldstein 2010; LaKind and Birnbaum 2010). Such prevention strategies could reduce the risk of disease in exposed populations, substantially reduce future mitigation costs occasioned by widespread environmental contamination, and encourage benign design and green-chemistry approaches to product development and waste disposal. Given that it will take many years to develop comprehensive toxicologic assessments for chemicals in commerce, exposure monitoring can help to ensure against unintended health and environmental consequences.

Customize Solutions

As stated in a 2009 National Research Council report (NRC 2009), the first step in a risk assessment should involve defining the scope of the assessment in the context of the decision that needs to be made. Adaptive exposure assessments could facilitate that approach by tailoring the level of detail to the problem that needs to be addressed. Such an assessment may take various forms, including very narrowly focused studies, assessments that evaluate exposures to multiple stressors to facilitate cumulative risk assessment, or assessments that focus on vulnerable or susceptible populations. Health-protective default values for exposure can be used in the absence of alternative information to expedite decision-making and to encourage generation of more chemical-specific data that are needed for risk assessment.

For example, a tiered approach that is customized according to expected future land uses and standardized health-protective default assumptions is incorporated into EPA's supplemental soil screening guidance (SSG) (EPA 2002). The SSG is a tool used to standardize and speed the assessment and cleanup of contaminated soil at sites on the National Priorities List (Superfund). The original 1996 EPA SSG (EPA 1996) focused exclusively on residential land use and did not incorporate alternative scenarios, such as future commercial or industrial land use (which would require less thorough cleanup), risks to workers, dermal exposure pathways, or inhalation from indoor vapor intrusion. The 2002 revisions allowed greater customization of the solutions according to the populations that would probably be exposed (for example, children and workers), anticipated exposure pathways, and future use of the land. The SSG is designed to be used only as a first-tier approach to evaluating a site, and more detailed and fully customized exposure assessments should be done if the SSG indicates a potential concern. This example illustrates that it is possible for agencies to develop standardized approaches to exposure assessment that can allow rapid assessment and conservation of resources while incorporating an increasingly customized and site-specific approach where needed.

Engage Stakeholders

Engaging broader audiences, including involving scientists in the concerns and needs of the public, will make the field more responsive and can improve problem formulation, monitoring and collection of data, access to data, and development of decision-making tools. Ultimately, the scientific results derived from the research will empower individuals, communities, and agencies in preventing and reducing exposures and in addressing environmental disparities. Engaging stakeholders will also mean moving beyond the science of inquiry to the science of engagement and the science of application (Boyer 1996). (Additional discussion on engaging the community to respond to health concerns is addressed in Chapter 6.)

Exposure science has played and will continue to play an important role in providing scientific data for use in assessing whether socially disadvantaged groups suffer disproportionate adverse exposures. Disadvantaged groups often have less access than other groups to beneficial or health-promoting exposures (such as access to parks and green spaces), an important consideration for population health (Sister et al. 2010; WHO 2010). And known interactions between psychosocial stressors and environmental exposures need to be considered and quantified for exposure science to be responsive (Shankardass et al. 2009).

Any effort to identify and quantify exposures in a way that fully addresses environmental disparities and that is capable of capturing the complex interactions between stressors in communities will need to use current measurement strategies and deploy new tools in exposure science broadly. Widespread implementation of biomonitoring of many chemicals in varied populations (on the basis of internal markers of exposure) will be useful in identifying exposure distributions and disparities. Key however is the need for exposure information to be accessible to community members and for them to have input in decisions involving exposure prevention or intervention. The committee maintains that it will be critically important to use such approaches as coupling of biomonitoring with collection of relevant environmental exposure data, source data, and health data to allow interpretation of the implications of exposures to facilitate prevention and intervention.

Participatory sensing techniques, which allow people to collect data on their own activities and on their communities, can be enabled for specific pollutants and exposure routes. However, people who are to be active in such programs need training, in both how to collect such data, and in its utility and limitations. Additionally the results need to be quality-assured and validated by trained professionals. For example, ubiquitous sensing technologies, such as those of smart cellular telephones, can facilitate collection of data on time–activity relationships that can provide information to support the need for measurements of pollutant exposures in many poorly characterized microenvironments. The development of user-friendly and less expensive monitoring equipment can allow trained people in communities to collect and upload their own data in partnership with researchers. Such partnerships would improve the value of the data collected and make more data available when setting priorities for exposure-control options. The collection and interpretation of such data raises many scientific questions and underscores the efforts needed to validate measurements and to determine how to integrate the information into models that support effective stakeholder engagement and decision-making.

Enhancing Exposure Science

Human and ecologic systems are inextricably linked. As part of the ecosystem, humans affect and are affected by interactions with the other organisms and nonliving components of the environment. At the most fundamental level, human health and well-being depend on the goods (for example, clean air and water) and services (for example, groundwater recharge, carbon sequestration, pollination, and seed dispersal) that are provided by ecosystems. Ecosystem health protection will require an expansion of problem-solving approaches that recognize that humans are an integral part of complex ecosystems that operate on wide spatiotemporal scales and varying levels of biologic organization, with consideration of exposures occurring both outside and inside exposed organisms.

The better we understand the complex interactions and feedbacks among the various human and nonhuman components of ecosystems, the better we will understand the spatial and temporal variability in risk and magnitude of exposure. Incorporating ecologic information into exposure assessments will allow us to identify how ecosystems cause, buffer, or magnify exposures. By broadening the view of receptors to include both ecologic and human receptors through the eco-exposome concept, exposure science will be able to connect stressors to changes in ecosystem function and in the ecologic goods and services on which society depends. From an operational perspective, the first step toward that integration will occur during the problem-formulation stage, where the human–ecology linkages can be articulated explicitly. In broadening their view of receptors, researchers and regulators will consider not only human health outcomes but, when it is feasible, ecosystem attributes and their interdependences.

Major challenges in exposure science, combined with the opportunities presented by new technologies, suggest the need for a transformation in exposure science. Strategic investments in this transformation are crucial for development of health-protective strategies in the 21st century. The investments must address strategies for research, education and training, and outreach for the development of collaborative and responsive frameworks for implementing these strategies in a resource-constrained environment. Moving forward with such a vision will provide a strong scientific basis for policy decisions that are responsive to a broad array of stakeholders.

REFERENCES

Boyer, E. 1996. The scholarship of engagement. Journal of Public Service and Outreach 1(1):11-20.

EPA (U.S. Environmental Protection Agency). 1996. Soil Screening Guidance: Technical Background Document. EPA/540/R95/128. Office of Solid Waste and Emergency Response, U.S. Environmental Protection Agency, Washington, DC [online]. Available: http://www.epa.gov/reg3hwmd/risk/human/rb-concentration_table/chemicals/SSG_nonrad_technical.pdf [June 19, 2012].

EPA (U.S. Environmental Protection Agency). 2002. Supplemental Guidance for Developing Soil Screening Levels for Superfund Sites. OSWER 9355.4-24. Solid Waste and Emergency Response, U.S. Environmental Protection Agency, Washington, DC [online]. Available: http://www.epa.gov/superfund/health/conmedia/soil/pdfs/ssg_main.pdf [June 19, 2012].

Goldstein, B.D. 2010. MTBE: A poster child for exposure assessment as central to effective TSCA reform. J. Expo. Sci. Environ. Epidemiol. 20(3):229-230.

LaKind, J.S., and L.S. Birnbaum. 2010. Out of the frying pan and out of the fire: The indispensable role of exposure science in avoiding risks from replacement chemicals. J. Expo. Sci. Environ. Epidemiol. 20(2):115-116.

NRC (National Research Council). 2009. Science and Decisions: Advancing Risk Assessment. Washington, DC: National Academies Press.

Shankardass, K., R. McConnell, M. Jerrett, J. Milam, J. Richardson, and K. Berhane. 2009. Parental stress increases the effect of traffic-related air pollution on childhood asthma incidence. Proc. Natl. Acad. Sci. USA 106(30):12406-12411.

Sister, C., J. Wolch, and J. Wilson. 2010. Got green? Addressing environmental justice in park provision. GeoJournal 75(3):229-248.

WHO (World Health Organization). 2010. Environment and Health Risks: A Review of the Influence and Effects of Social Inequalities. World Health Organization [online]. Available: http://www.euro.who.int/__data/assets/pdf_file/0003/78069/E93670.pdf [accessed Mar. 26, 2012].

3

Applications of Exposure Science

INTRODUCTION

Exposure science plays a fundamental role in the development and application of epidemiology, toxicology, and risk assessment. It provides critical information for protecting human and ecosystem health. Exposure science also has the ability to play an effective role in other fields, including environmental regulation, urban and ecosystem planning, and disaster management; in many cases these are untapped opportunities. Exposure science links human and ecologic behavior to environmental processes in such a way that the information generated can be used to mitigate or prevent future adverse exposures. This chapter discusses current and future opportunities for and challenges to applying exposure science to those fields and frames some of the needs for further development of exposure science. Text boxes are intended to illustrate specific examples of the role of exposure science.

EPIDEMIOLOGY

Exposure assessment is a major component of environmental epidemiology; it is equal in importance to outcome assessment. Historically, the main focus of many environmental epidemiology studies has been on single chemical, biologic, and physical stressors (for example, individual pesticides, air pollutants, ionizing radiation, or water contaminants). But human populations are exposed to a multitude of potentially hazardous stressors simultaneously, often with highly correlated patterns of spatial and temporal variation, and are affected by human activities and behaviors, so it is difficult to pinpoint the specific stressor associated with a particular outcome. New high-throughput genomic and biomonitoring technologies (discussed in Chapter 5) are providing for a greater number of potential biomarkers that can be used to assess multiple exposures simultaneously. In addition to chemical, biologic, or physical stressors, epidemiologists may be concerned with psychosocial stressors that influence

Applications of Exposure Science

disease risks directly or modify a person's susceptibility to the effects of other agents (Shankardass et al. 2009). The direct effects of psychosocial risk factors on health outcomes are beyond the scope of this report, but their role as modifiers of exposures is within the committee's charge. Figure 3-1 provides a general schema for thinking about the components of exposure assessment in environmental epidemiology. Each of the components shown in solid boxes represent areas where exposure science can contribute, providing data on determinants of people's exposures (the intensity of the spatio-temporal exposure fields they are moving through and their personal time/activity patterns in that field) and direct measurements of individuals' external exposures or internal doses, along with relevant modifying factors. The hypothesized (but not directly observable) history of doses to the relevant target organs are then used by epidemiologists to relate to the health outcomes.

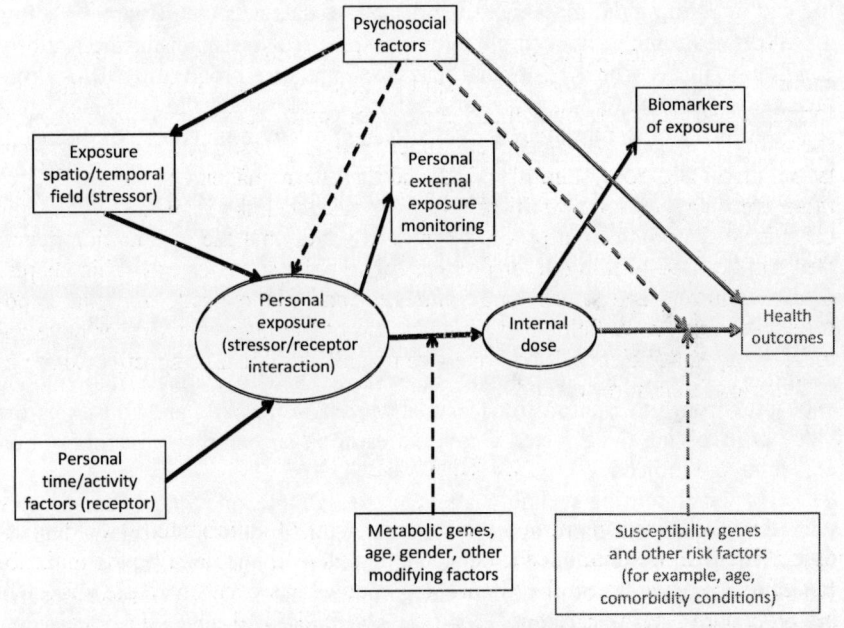

FIGURE 3-1 General schema of exposure assessment in environmental epidemiology. Items in gray, related to health outcomes and their determinants other than environmental exposures, are included to place exposure assessment in context but are outside the charge of this committee. Boxes represent measurable quantities, and ovals denote hypothetical intermediate variables that can be assessed only indirectly. Solid arrows denote direct effects, and dashed arrows indicate modifying effects. For example, the oval marked "personal exposure" represents the entire history of an individual's true biologically-relevant exposure, which is not directly observable. The boxes labeled "personal external exposure monitoring" and "biomarkers of exposure" represent the data that are potentially observable.

Traditional exposure assessment for epidemiology combines the use of measurements and models to characterize the spatio-temporal field of environmental concentrations of a stressor with individual data on interactions of receptors (people) with their environment (for example, derived from questionnaires on time–activity patterns) to estimate personal exposures. In some studies, direct measurements of personal exposures (for example, film-badge measurements of occupational radiation exposures) and novel methods of tracking individual activities (for example, Global Positioning System (GPS) monitoring of locations and accelerometers for physical activity rates and videotaping of activities) have been used, either on an entire study population or on some sample for calibration or validation of model predictions. The emerging field of molecular epidemiology, based on the use of biomarkers of exposure (as well as of susceptibility or early signs of disease), offers potentially transformative advances in exposure science, particularly if combined with novel genomic, transcriptomic, metabolomic, and other "–omic" technologies and bioinformatic tools for organizing and integrating the massive, often disparate, data sets (see Chapter 5). Box 3-1 illustrates some of the complexities of exposure assessment for the National Children's Study, with longitudinal measurements of a broad array of environmental and personal (external and internal) exposures and health outcomes.

Historically, comprehensive *measurement* of environmental exposures has not been possible, requiring statistical models to interpolate among relatively sparse measurements. The models can be purely statistical, such as geostatistical models for air pollution, or can be based on mathematical models for tracking agents from sources through intake by receptors (see NCRP 2010a for a discussion of the general principles of environmental dose reconstruction for radiation exposures and NCRP 2008, 2010b for recommended approaches to uncertainty analysis for external and internal exposures respectively). Box 3-2 provides an example of environmental pathway analysis applied to evaluate radionuclide exposures from the Hanford nuclear plant in Hanford, WA, and illustrates the value of involving the affected communities in all stages of the planning of an epidemiology project.

As novel sensing technologies, such as satellite imaging, become more widely available and more accurate, the need for models will remain, but the focus will shift from interpolation to exploitation of massive datasets. A key function of models is not just to provide point estimates of individual exposures but to quantify the uncertainty in exposure estimates, to understand measurement error in health analyses.

Environmental exposures typically occur over extended periods of time at varying intensities, requiring a shift in thinking from simple exposure-response to exposure-time-response relationships (Thomas 1988). These can be quite complex, involving modifying effects of age-at-exposure (for example, at particularly sensitive developmental stages), time-since-exposure, duration-of-exposure, or other time-related factors. In addition, for most conditions, little is known about whether short intense exposures have larger or smaller effects than

> **BOX 3-1** Case Study of Exposure Assessment for
> the National Children's Study
>
> The National Children's Study (NCS) is a nationwide cohort study of pregnancy outcomes and child development through the age of 21 years. The study aims to investigate the separate and combined effects of environmental exposures on pregnancy outcomes, child health and development, and origins of adult disease. Environmental exposures in the NCS are broadly defined to include chemical, biologic, physical, psychosocial, and genomic factors. The study also aims to examine determinants of child, maternal, and developmental health disparities, such as prenatal exposures, geography, social status, race, ethnicity, neighborhood characteristics, and quality of social networks, and the impact of various disparities on health outcomes. The current study has about 100 study locations, but the number, size, and selection process for these study locations may evolve (Hirschfeld et al. 2011).
>
> A complex exposure assessment is planned, entailing a combination of techniques—questionnaires on diet and product use; environmental sampling of house dusts; and collection, storage, and assays of biologic specimens (for example, blood, urine, saliva, hair, breast milk, and adipose tissue) (Barr et al. 2005; Needham et al. 2005). Depending on the particular agent, exposure assessments will be conducted at various life stages, for example, in utero, postnatal, and peripubertal (Bradman and Whyatt 2005; Wang et al. 2005; Landrigan et al. 2006). In addition, data from national and state agencies are being used to profile areas within a county and locations of the study participants (Lioy et al. 2009; Downs et al. 2010); a test case is being conducted in Queens County, NY (Lioy et al. 2009). Sophisticated modeling will be needed to combine the various data sources.
>
> Given the size, scope, and complexity of the NCS, there have been challenges to identify exposure assessment approaches and methods that are feasible, acceptable, and limited in both cost and participant burden. Validation sub-studies have been investigated (Strauss et al. 2010), extant data will be relied on, and choices will have to be made about which chemicals to measure in stored environmental or biologic samples (Özkaynak et al. 2005; Gilliland et al. 2005; NRC/IOM 2008).
>
> The development of the exposure assessment component of the NCS study highlights the challenges for exposure science to meet the demands for exposure information across time scales for large populations.
>
> In 2010, EPA, NIEHS, and the NCS organized a workshop to engage scientists from the exposure, epidemiology, and health effects disciplines with the goal of identifying the most promising and practical exposure metrics to use in a study the size and scope of the NCS (Tulve et al. 2010). For the most part, the workshop participants agreed that questionnaires and diaries provide inadequate and unreliable exposure information; that more effective exposure metrics are needed that will provide better information on exposures and their inter- and intra-individual variabilities, and that high quality samples should be collected (in particular in the perinatal period) and archived for future analyses as new analytic methods are developed.

> **BOX 3-2** Case Study of the Hanford Environmental Dose-Reconstruction Project
>
> As part of the Manhattan Project and continuing into the 1960s, all the plutonium for the U.S. nuclear-weapons program was produced at the Hanford Nuclear Reservation in the southeastern part of Washington State. During the early years, considerable quantities of radionuclides were released into the environment, notably iodine-131, which tends to accumulate in the thyroid gland of exposed people and can lead to thyroid cancers and other thyroid abnormalities. To address concerns of the downwind population, two projects were launched by the Centers for Disease Control and Prevention: the Hanford Environmental Dose Reconstruction (HEDR) Project and the Hanford Thyroid Disease Study (HTDS). The latter was an epidemiologic cohort study that made use of dose estimates produced by the HEDR Project. The HEDR Project is an important illustration of how exposure science informs epidemiologic research.
>
> The HEDR Project began by reviewing over 38,000 pages of environmental-monitoring documents. Technical panels of experts in nuclear engineering, radiation dosimetry, environmental transport, meteorology, hydrology, statistics, and other fields developed an environmental-pathway model. The model reconstructed the releases from the plant and modeled their transport through air, soil, and water contamination; uptake by vegetation; intake by dairy cows and goats; milk production and distribution and ingestion by individual study participants and their mothers while pregnant; and ultimately dose delivery to the thyroid gland. Some parts of the complex model were informed by measurements, others by expert judgment; each step entailed careful consideration of the relevant uncertainties. The entire model was incarnated in a Monte Carlo computer program that yielded multiple realizations of possible doses to each individual, with variability among the realizations reflecting the uncertainties in the final dose assignments (Shipler et al. 1996). Although much concern had been expressed about contamination of the Columbia River, the analysis found the contribution to human exposures by that pathway to be negligible. The distribution of final dose estimates used by the HTDS had a range that was shown to provide adequate power to test for dose–response relationships with various thyroid outcomes (Kopecky et al. 2004). The null results for all types of thyroid abnormalities provided evidence that most cases in the region were unlikely to be due to radioactive releases from the Hanford plant (Kopecky et al. 2005), although this interpretation remains somewhat controversial (Hoffman et al. 2007).
>
> A major lesson yielded by the HEDR Project and HTDS was the importance of community involvement, particularly in light of a concurrent class action lawsuit and suspicions that the U.S. Department of Energy (the owner and operator of the site) was influencing the study. Public, state, and American Indian representation in the independent Technical Steering Panel (TSP) that oversaw the project was viewed as essential from the beginning (Shipler 1995). The TSP adopted a commendable policy of openness in all aspects of
>
> *(Continued)*

> **BOX 3-2** Continued
>
> the project, including definition of scope, budget, and priorities. The investigators commented that the open approach required more time and energy but concluded that "if these issues and concerns are addressed early and if scoping of the project with an advisory panel is completed before major work is initiated, cost-effective planning and management can be achieved and 'science in a fishbowl' can be successful" (Shipler 1995, p. 108).

long low-intensity exposures or other patterns of temporal variation. While such issues are amenable to epidemiologic analysis, this is possible only if detailed histories of exposure are available, requiring exposure scientists to develop novel ways of reconstructing the entire history of past exposures or for monitoring time-varying exposures prospectively for extended periods.

The "exposome" concept (see Chapter 1) may provide a framework for representing a person's lifetime of exposure to all potentially hazardous or beneficial agents. Although the concept is generally interpreted as relating to the totality of biologically-relevant exposures of either external or endogenous origins, the current enthusiasm is driven largely by the rapid advances of various –omics technologies (reviewed in Chapter 5) that permit agnostic assessment of a broad swath of internal biomarkers of exposure. For example, two recent publications illustrate its potential utility with "Environment-Wide Association Studies (EWAS)," agnostic scans for associations using a panel of a few hundred metabolite measurements (Box 3-3). These illustrations are analogous—albeit on a smaller scale—to genome-wide association studies that test the association of a disease or trait with hundreds of thousands to millions of genetic variants, but provide a "proof of concept" for an approach that could in principle be extended to a much broader range of exposures, monitored longitudinally. Novel approaches are needed to mine such data (Thomas 2010), together with internal and external markers of exposure to improve assessment of exposure-response relationships and, more importantly, to find ways to intervene before an adverse outcome is observed in an individual or population. That is a long-term goal that will require new approaches for conducting research, including capitalizing on future advances in individualized medicine and understanding the effects of changes in lifestyle and human behaviors.

Exposure assessment is usually constrained by cost or other feasibility considerations. It is seldom possible to measure everything that one would like to measure over the lifetime of an entire epidemiologic cohort. Hence, various sub-study designs are needed to devise a cost-efficient method of exposure assessment. That typically entails statistical modeling to combine the substudy data with the main-study data (Breslow et al. 2009). The study-design challenge involves trying to optimize the various tradeoffs—for example, between numbers of subjects and locations and breadth and duration of measurements—to yield the most precise estimates of the exposure–response relationship of interest. Key to the optimization challenge is the measurement-error distribution expected with alternative designs and the use of statistical methods for adjusting for measurement errors (Carroll et al. 2006). Another type of hybrid design entails combining individual and group measurements. For example, one might correlate disease rates in large populations with estimates of the joint distribution of exposure, confounders, and modifiers obtained from sample surveys within each population (Sheppard et al. 1996). Box 3-4 illustrates the value of improved exposure estimates for epidemiologic studies.

The availability of various population-wide outcome databases—such as databases of mortality, hospitalization, and cancer incidence—is unfortunately not matched by population-wide exposure databases. The availability of a national dose registry for radiation workers in Canada, however, has made it possible to use record-linkage techniques to conduct large-scale studies of dose–response relationships for cancer (Zablotska et al. 2004). Establishing such registries and extending them to include medical-radiation doses, perhaps in the form of an electronic personal dose history, would be a boon for the field of radiation epidemiology. Ultimately, it would be desirable to have some life-course exposure registry for the entire population, or a periodic census of a large sample of the population that would inquire about a broad spectrum of environmental exposures for research purposes.

BOX 3-3 Environment-Wide Association Study (EWAS)

Patel et al. (2010) conducted an agnostic scan for associations of many measurements of internal exposures with type 2 diabetes; this was similar in spirit to Genome-Wide Association Studies (GWAS) that test the association of a disease or trait with hundreds of thousands to millions of genetic variants. Rather than using traditional methods of characterizing external exposures, the investigators used the "exposome" concept to assess potentially biologically effective exposures with a panel of 266 metabolite measurements obtained from the National Health and Nutrition Examination Survey of 503–3,318 people. They found statistically significant associations (after adjusting for multiple comparisons) with heptachlor epoxide (a pesticide derived metabolite), vitamin γ tocopherol, and some PCBs and found protective effects of β carotenes. A similar EWAS (Patel and Butte 2010) looked at associations with gene expression levels. Because it was not a longitudinal study, there is the potential for "reverse causation" bias, a tendency for disease or its treatment to affect biologic measurements rather than for the exposure to be a cause of the disease. Cohort studies would avoid that difficulty by relating biomarker measurements in unaffected people to their later onset of new disease. The Patel et al. study should be considered as a "proof of concept" for an approach that shows great promise for application of far more extensive panels of biomarkers from the various biobanks being assembled or already in existence that have stored biologic specimens from hundreds of thousands or millions of subjects with followup for disease incidence.

> **BOX 3-4** Value of Improved Exposure Estimates for Epidemiologic Studies
>
> Reducing exposure error is critical for epidemiologic investigations. Previous air-pollution health-effects studies have underlined the importance of capturing spatial variability, particularly in urban areas (Logue et al. 2010; Bell et al. 2011). Accurate assessment of human exposures to atmospheric pollution requires knowledge of the spatial distribution of pollutants over cities on scales of 1-100 m (Chow et al. 2002). The improved resolution is expected not only to reduce exposure-assessment error but generally to result in larger health-effects estimates. For example, Jerrett et al. (2005) applied kriging techniques to study the association between within-city $PM_{2.5}$ exposure gradients and mortality and found a substantially larger effect than previously reported with the city-average exposures (Liu et al. 2009, p. 886). The Women's Health Initiative found a larger pollution effect on mortality when within-city exposure estimates were used (Miller et al. 2007). Nevertheless, although it is generally true that reductions in exposure measurement error can lead to improvements in health-effect estimates, it is not always true, and it depends on the specifics of the measurement error and the true exposure distributions (Szpiro et al. 2011). (See discussion in Chapter 5.)

The goal of a truly population-based exposure registry may be less feasible in the United States than in countries that have national health systems and population registries, at least in the foreseeable future. However, health-maintenance organizations (HMOs) may provide unique opportunities to build large-scale databases that, when combined with biomarkers of exposure assayed from routinely collected biospecimens and systematically collected exposure information (from clinic visits or questionnaires), could form the basis of long-term cohort studies. Outcome data reflecting clinic visits, hospitalizations, diagnoses, medication prescriptions, and mortality would be routinely available through followup data collection. Although not strictly random, the coverage of the larger HMOs is extensive enough to represent a broad spectrum of the population. Exposure science could also take advantage of data obtained on individuals and populations through market-based and product-use research to improve questions on exposures in epidemiologic studies.

TOXICOLOGY

Toxicology, whether focused on mechanisms or hazards, has historically been conducted outside the context of actual human exposures. Two major hazard evaluation programs, the U.S. Environmental Protection Agency (EPA) ToxCast and the National Institute of Environmental Health Sciences National Toxicology Program select chemicals and other materials (such as nanomaterials) for evaluation and use exposure as one qualitative selection criterion (Dix et al. 2007). But exposure context is for more than selection of chemicals for testing. The biology of systems perturbed by exposures to stressors is highly sensitive to the magnitude of exposure (Slikker et al. 2004; Andersen et al. 2010). Mechanistic studies and hazard assessments conducted at concentrations that far exceed actual human exposures may produce results that are misleading because the observed effects are not likely to occur at lower doses or because low-dose effects may be masked by more overt toxicity at high doses. The availability of more exposure data could guide dose concentrations in toxicity studies.

The interdependence of toxicology and exposure science is recognized in these two communities, but exposure science is typically underemphasized as a principle in toxicology. The committee responsible for *Toxicity Testing in the 21st Century: A Vision and A Strategy* (NRC 2007) recognized the need for better integration and use of exposure science in toxicity assessment and called for its greater use in each step of the vision. The report spurred the rapid development of toxicity testing, in particular in in vitro, high-throughput methods, but its use of exposure science has seen little growth over the same period. That is unfortunate in light of the fundamental interdependence of the two fields, including the importance of exposure information in the design and interpretation of toxicity testing (Cohen Hubal et al. 2010; 2011). In place of current practice, the present committee envisions a shift toward a toxicologic assessment program that has an interface with exposure science and is influenced by and responsive to human and environmental exposure data. Such a program would strengthen the current toxicology-driven paradigm by focusing on the four activities described below.

- *Select and set priorities among chemicals for toxicity testing.* As EPA implements the recommendations of the 2007 National Research Council report, exposure science will become even more important for priority-setting (Cohen Hubal et al. 2010). As described in Chapter 2, that report envisioned a process for screening chemicals in commerce for hazard potential with rapid toxicity-pathway screens informed by and with priorities set through screening-level exposure assessments. The EPA ToxCast program is one example of such an implementation effort. In addition, efforts to reform the Toxic Substances Control Act (TSCA) will probably rely on priority-setting strategies that consider both exposure and toxicity potential. For those reasons, the present committee's vision of enhancing the publicly available information on chemicals in commerce, improving screening for chemicals in the environment and in people (via biomonitoring and microsensor networks), and improving exposure modeling will form a solid foundation for priority-setting in relation to toxicity-pathway screening studies.
- *Provide internal and external exposure information to inform selection of relevant concentrations of stressors for high-throughput toxicity testing.* Exposure science needs to develop strategies to provide the information required to enable testing of stressors in animals. However, statistical-power considerations may make it infeasible or inadvisable to use only environmentally relevant doses in whole-animal studies. Newer in vitro toxicity-pathway studies can be con-

ducted at a wide range of doses, and these studies can be informed by exposure-related information, especially as it relates to internal dose. Reaching that goal will require a shift in exposure science toward collection of internal measures of exposure, as discussed in Chapter 5.

- *Provide quantitative pharmacokinetic data (on absorption, distribution, metabolism, and excretion) derived from human-exposure studies.* Targeted exposure studies need to include collection of exposure information to allow inference about human pharmacokinetic measures, such as the time course of exposure, for some high-priority chemicals. Exposure-characterization protocols should include measurements of external and internal markers of exposure to assess bioavailability, especially when exposures are predominantly via a single route. Greater use of longitudinal internal exposure studies that include periods of high and low or no exposures (such as those in occupational environments during and after work) could provide concentration time-course data similar to repeated-dose pharmacokinetic studies. For example, absorption rates, half-lives, and other pharmacokinetic measures and their variability within and between individuals could be derived from those data and would provide a wealth of critical human pharmacokinetic data for setting exposure concentrations for toxicity testing and for use in risk assessment (Teeguarden et al. 2011).

- *Link exposure data with in vivo data on perturbations of toxicity pathways in human or wildlife populations to identify exposure-response relationships directly.* The conventional hazard-assessment paradigm uses cell-culture systems or animal models to identify hazards. In the future, the present committee expects collection of higher-resolution and larger quantities of exposure data in a broader swath of the population to allow epidemiologists to identify potential hazards in human populations or ecosystems. The characterization of the hazards could then be explored by using more focused, efficient toxicologic studies at relevant exposure concentrations and durations or measurements of perturbations of toxicity pathways (that is, as seen in genetic or other biomarkers of effect) in exposed human or wildlife populations. Although these studies might be expected initially to focus on individual stressors, they would evolve with advances in exposure technologies to identify and characterize combinations of stressors.

There are also opportunities for epidemiology and toxicology to be more closely tied to the process of exposure assessment. For example, cell lines derived from epidemiologic study subjects could be exposed in culture to mixtures derived from samples from the specific environments to which the subjects were exposed, to identify measures of biologic activity of these complex mixtures. The measures of biologic activity would then represent more relevant, more specific measures of response for use as variables in subsequent epidemiologic studies. Exposure of experimental animal models to environmental mixtures associated with specific epidemiologic studies would provide additional information on exposure-response relationships, the time-course of development of disease, and the role of genetics as modifiers of exposure and response. In vivo challenge studies in humans or experimental animal models can also provide much information on intermediate biologic responses to agents (for example, diesel-exhaust particles or second-hand smoke) and on genetic modifiers of these responses to exposures (Gilliland et al. 2004). That information could inform exposure–response or gene–environment interaction analyses of epidemiologic data. Finally, epidemiology and toxicology would benefit from a more sophisticated approach to modeling that takes advantage of the source-to-effect continuum and reduces uncertainties in study design and misinterpretation of results for use in mitigation and prevention (Georgopoulos et al. 2009).

ENVIRONMENTAL REGULATION

Risk Assessment

Exposure assessment is one of the core components of regulatory quantitative risk assessment; therefore, the quality of exposure information and the state of exposure science are paramount in the quality and utility of risk assessment. The National Research Council and EPA have previously described the major steps in risk assessment, including hazard identification, dose–response assessment, exposure assessment, and risk characterization (NRC 1983, 1994, 2009). Although exposure assessment is often described as perhaps the most challenging component of risk assessment, prior National Research Council reports on risk assessment have made limited recommendations for improving the quality of exposure data or the utility of exposure assessment for quantitative risk assessment. Although those steps are important, a strategy for improving the quality and quantity of exposure data is needed to reduce uncertainties stemming from the exposure-assessment component of risk assessment.

A recent EPA Science Advisory Board panel provided nearly 100 recommendations for improving guidance in ecologic risk assessment (EPA SAB 2007; Dale et al. 2008) and suggested that further consideration was needed for assessing simultaneous exposure to multiple stressors, assessing spatial and temporal variation in exposures, and addressing uncertainties in exposure models.

Exposure assessment poses numerous challenges for risk assessment. Exposures change, so a risk assessment that uses data that are available today may no longer be valid months or years from now; this is especially true for chemicals newly entering the market, for which use and exposure patterns have not yet fully emerged. Important exposure pathways may be missed, and this can lead to underestimation of overall exposure or neglect of highly exposed populations. Risk assessments and exposure assessments tend to focus on one chemical at a time and potentially miss interactive effects that could influence both exposure and risk. Data needed for exposure assessment, such as data on chemical sales and product ingredients, may be proprietary and not publicly available because of trade-secrecy protections.

Applications of Exposure Science 61

To be useful for risk assessment, an exposure assessment needs to be capable of identifying and quantifying the exposure of the populations that are most highly exposed and populations that are most vulnerable. The assessment should strive to include all relevant exposure pathways and allow the pathways to be identified and defined individually (for example, to allow for quantification of exposure from water or from food as well as total exposure). Exposure assessment also needs to consider background exposures to chemical or nonchemical hazards that could influence cumulative exposure or risk within a population. A mechanism for tracking and updating exposure estimates is needed to ensure that they continue to reflect real-world conditions. Finally, uncertainty, which is inherent in an exposure assessment, needs to be quantified to determine the level of confidence in the overall risk assessment.

The 2009 National Research Council report *Science and Decisions: Advancing Risk Assessment*, which focused on improving human-health risk assessment but also considered ecologic risk assessment, contains several recommendations related to exposure assessment. It describes major challenges that risk assessment is facing and concludes that regulatory risk assessment has become bogged down and that many assessments take a decade or longer to complete. Strategies that the present committee believes may improve the efficiency, quality, and utility of the exposure-assessment component of risk assessment include the following:

- *Determine in advance the level of detail and type of exposure information needed to address the risk-management question at hand.* Risk assessment should be viewed as a method for evaluating the relative merits of various options for managing risk rather than as an end in itself (NRC 2009). It is important to pose the policy question first and then gather the information needed to answer it. In the case of exposure assessment, screening-level information may be adequate to address some questions, targeted data may be useful for others, and extensive data may be needed in some circumstances, particularly for developing scientifically sound policies and regulations.

- *Use health-protective exposure default assumptions when adequate data are not available and define the criteria needed to depart from them.* Default assumptions are typically used in risk assessments when measured or modeled data are unavailable or to allow extrapolation from existing data. Defaults are intended for a risk assessment to move forward in a timely fashion despite inevitable data gaps and to be health-protective. For example, one standard default assumption in exposure assessment is the choice of a cutoff for estimating exposure in a population (for example, the 90th or 99th percentile) (EPA 1992; NRC 2009). An exposure assessment used for risk assessment should select a cutoff that is as health-protective as the data allow and should explain the choice of the cutoff in the context of EPA's default assumptions and the existing data.

- *Quantify population vulnerability better.* An important purpose of exposure assessment is to identify populations that are at greater risk than the majority of the population because of their exposure patterns or susceptibility to health effects. For example, for a risk assessment of methylmercury, it is essential to gather data on people who fish for subsistence, because these people are expected to face the highest exposures (EPA 2000). In the example above, instead of selecting a population cutoff (such as the 99th percentile), a better approach would involve identifying the subpopulation in the upper end of the exposure distribution, measuring exposure in that subpopulation, and identifying strategies to reduce the upper-end exposures.

- *Assess and quantify cumulative and aggregate exposures.* Exposures may involve aggregate exposures to a given chemical via a variety of exposure pathways or may involve cumulative exposures to various chemical or nonchemical stressors that influence human exposure and health risk by influencing metabolism or excretion of a chemical and acting on the same biologic pathway or affecting the same organ system. There is a need to include multiple chemical, physical, or biologic stressors but also to consider other vulnerability and susceptibility factors that influence the effects of these stressors, such as nutritional and psychosocial status, including stress. Exposure assessments may therefore need to collect information about exposure to a variety of chemical and nonchemical stressors that may interact to influence health risk.

- *Improve stakeholder involvement and make the process more accessible to the general public.* Exposure information used as the basis of a risk assessment needs to be made available to the public and presented to enhance understanding, particularly regarding the assumptions used and the uncertainties and limitations of the data. Efforts should be made to make all models used to estimate human exposure for risk-assessment purposes nonproprietary and open to public scrutiny so that the basis of modeled data is clear and transparent. In addition, an open-source strategy for modeling software and algorithms could ensure an ongoing process of peer review and improvements in model specifications while ensuring transparency (for example, PLOTS 2012). Information on chemical use and product ingredients also needs to be made publicly available to the greatest extent possible to allow the public to understand exposure pathways and to ensure that important exposure information is not hidden by trade-secrecy provisions.

Box 3-5 illustrates the role that exposure data and risk assessment have played in the establishment of a drinking-water standard for perchlorate.

Risk Management

Exposure science can inform public policy in a broad array of scenarios. Exposure information, even with minimal or no data on chemical hazard, can provide important information about trends, disparities in population, geographic "hotspots," cumulative exposures, and predictors of vulnerability. Fur-

> **BOX 3-5** Case Study of Perchlorate in Drinking Water
>
> In 1997, perchlorate, a chemical component of rocket fuel was detected in several drinking water wells in Sacramento County, CA, near a large industrial facility. In response, the California Department of Public Health laboratory developed a new analytic method for detecting perchlorate at lower concentrations than was previously feasible—down to 4 ppb. The method was later used to conduct a survey of multiple drinking water and agricultural wells in California, including in densely populated Los Angeles, Riverside, and San Bernardino Counties, where perchlorate was detected in numerous wells and nearby pollution sources were identified. Sampling also showed perchlorate at low concentrations (5-9 ppb) in Colorado River water, an important source of drinking water and water for agriculture in southern California (CDPH 2010a).
>
> The widespread detection of perchlorate in drinking water catapulted it to public attention and drove action in numerous states to develop enforceable drinking water standards. Further discoveries of the chemical in agricultural wells spurred extensive investigations of perchlorate uptake into food by the Food and Drug Administration and by independent scientists, and the resulting information was used in exposure assessments by the California Environmental Protection Agency and others to derive risk based standards. Development of a more sensitive analytic method for detecting perchlorate in water supplies ultimately drove the cleanup of numerous industrial and military sites and the creation of a National Research Council committee that reviewed the U.S. EPA's risk assessment for perchlorate. EPA cited the widespread exposure to this chemical as a major justification for their recent decision to promulgate a national drinking water standard for perchlorate.
>
> It is apparent from this case study that environmental monitoring of exposure sources can identify significant public health issues, inform science and risk assessment, and drive risk management.

thermore, the very existence of exposure data creates an imperative to generate information about hazard so that the importance of the exposure can be better understood. Several important existing and potential applications of exposure science in public policy are described below.

Exposure-based prevention. A screening-level exposure assessment can inform decisions before new chemicals enter the market. Currently, that is done in the context of a formal risk assessment of pesticides. For new industrial chemicals on which exposure data are not available, EPA considers projected use and predicted chemical properties in making a decision about whether to request testing under TSCA, although in reality this is rarely done (GAO 2009). Screening exposure assessments could be used much more aggressively. For example, consideration could be given to keeping chemicals that are highly likely to persist and bioaccumulate off the market, so as to avoid the potential to contaminate the environment at increasing concentrations over time and potentially result in human and environmental hazard. Another example of exposure-based prevention would involve identification of chemicals that are predicted to contaminate water because of their persistence and solubility and restriction of uses that could threaten surface water or groundwater. Finally, testing of consumer products in premarket controlled studies can be used to prevent toxic materials from reaching humans, and causing harmful exposures.

Exposure-based control. Post-marketing surveillance of chemical exposures can be used to ensure safety with respect to chemicals used or distributed in large quantities. About 200 chemicals were included in the most recent *National Report on Human Exposure to Environmental Chemicals* (CDC 2011). Other chemicals are studied by individual researchers or in state programs. There is no requirement that companies conduct post-marketing surveillance, and the Centers for Disease Control and Prevention does not have the resources to screen routinely for the thousands of chemicals that are used and distributed in large quantities or to which people may be exposed. Broad exposure surveillance of a longer list of high-priority chemicals would provide important information about trends, exposure variability, and magnitudes of exposure. Such surveillance would help evaluate whether regulatory actions designed to control exposure are effective, would allow priority-setting among chemical exposures to which are increasing, and would target at-risk populations for exposure reduction. For chemicals that can be biomonitored, surveillance could initially focus on worker populations and ecosystems near manufacturing or processing facilities. For widely dispersed chemicals (for example, those in consumer products), exposure of the general public or specific populations could be monitored. For chemicals that cannot be biomonitored (for example, because they have short biologic half-lives), near-source environmental monitoring or modeling of exposure could be possible. If exposure of any population is above a screening level of concern or if exposures are shown to be increasing, policy action could reduce exposure to prevent potential health problems from developing. Examining product-use patterns and the release of contaminant residual from products can keep highly toxic contaminants from coming into contact with consumers or can help to reduce the effects of such exposures.

Exposure justice and equity. The reduction of disparities in exposure to chemical and nonchemical stressors is an important goal of public policy. That principle is reflected in Presidential Executive Order 12898 (February 11, 1994), which calls on federal agencies to "improve research and data collection relating to the health of and environment of minority populations and low-income populations (p. 2)", including "multiple and cumulative exposures" (p. 3). The Executive Order expects, furthermore, that "each Federal agency shall make achieving environmental justice part of its mission" (p. 1).

For agents on which there are population exposure data, exposure disparities can be tracked by evaluating the distribution of the exposure concentrations over time (for example, the median, mean, standard deviation, or range).[1] It is important that exposure science focus on identifying the segments of the population that are in the top end of the exposure range for any given stressor. These "high-end" exposed populations need to be explicitly assessed so that the exposure pathways can be understood and addressed. Policy actions could appropriately include engagement of the affected communities in a collaborative process to identify and reduce the sources of exposure (for example, Brown et al. 2012). Community participation can help ensure that the research questions that are being asked are relevant to the needs and concerns of both the researchers and the affected community, increasing the likelihood that the project will contribute to improving public health (O'Fallon and Dearry 2002). Progress toward that goal will be seen in a reduction in the skewness of the population exposure data or through followup assessment in targeted populations.

In addition to population-based surveillance to reduce exposure disparities, exposure science has an important role in performing rapid on-site assessments of the relationship between potential exposure and potential health outcomes in communities. Such assessments have been done around hazardous waste sites and in other settings where contaminants have already been introduced into the environment, but can also be used more consistently as a rapid exposure screening tool before contaminants are inadvertently introduced into communities or the personal environment.

Box 3-6 illustrates the role that exposure science played in managing potential risks posed by polybrominated diphenyl ethers (PBDEs), including spurring the creation of the California biomonitoring program and encouraging a broad strategy for identifying emerging chemicals of concern in consumer products.

Compensation Policy

Exposure science plays an important role in policy decisions regarding compensation of people who may have been harmed by exposure to hazardous substances. Many government programs have been established to provide compensation to veterans exposed to various deployment-related or combat-related exposures (for example, Agent Orange exposure of Vietnam veterans), to workers (for example, uranium miners), or to members of the general public (for example, people downwind of the Nevada nuclear test site). In addition, state workers' compensation programs often have substantial difficulty addressing questions of causation of potential occupational illness due—at least in significant part—to a lack of exposure data for workers.

---

[1]Extremely skewed data can be identified easily because the mean is significantly higher than the median, and the standard deviation is large. Although most exposure data do exhibit a log-normal (skewed) pattern, the degree of skewness varies.

> **BOX 3-6** Case Study of Chemicals in Breast Milk:
> Policy Action Based on Exposure Data
>
> In 1999, researchers in Sweden discovered a previously unknown group of chemicals in women's breast milk. When the researchers tested stored breast-milk samples from the 1970s and 1980s, they discovered that the chemicals had increased dramatically over the decades, doubling every 5 years (Meironyte et al. 1999). Soon after the Swedish discovery, U.S. scientists reported the same chemicals in breast fat and breast milk in Indiana, Texas, and California (Schecter et al. 2003). The U.S. levels were much higher than those reported in Europe; the California samples had concentrations 40 times higher than those in Sweden. The chemicals, polybrominated diphenylethers (PBDEs), were flame retardants commonly used in consumer products. At that time, the PBDEs were essentially untested for toxicity to humans or ecosystems, but they were in widespread use in foam cushions, fabrics, and electronics. PBDEs are structurally similar to polychlorinated biphenyls and dioxins, and they share environmental chemical characteristics of persistence and bioaccumulation.
>
> Those discoveries led to numerous policy actions: PBDEs immediately had high environmental-health research priority, sparking numerous exposure, toxicology, and epidemiology studies; the European Union banned most of the PBDEs in 2004; California and other states passed legislation to phase out some of or all the PBDEs; and EPA reached an agreement with the U.S. manufacturer to cease production of the most bioaccumulative PBDE voluntarily, and other PBDEs were later phased out of U.S. production. Although these chemicals are still being produced in China and other countries. The California statewide biomonitoring program was created by the legislature in large part because of the discovery of PBDEs in breast milk and breast tissue. The legislation identifies one of the priorities of the program as "the need to assess the efficacy of public health actions to reduce exposure to a chemical" (CDPH 2010b, p. 32). The California Scientific Guidance Panel added an additional criterion that was based on the need for early identification of emerging chemicals of potential concern (CDPH 2010b). One of the first actions of the program was to list as priority chemicals "chlorinated and brominated organic chemicals used as flame retardants" (CDPH 2010b, p. 6). The broad language of the listing was designed to address the chemicals of concern that were increasing in market share to replace the PBDEs.

Some instances of determination of compensation, such as "presumptive disability policies" for some classes of veterans intended to give the veterans the "benefit of the doubt", require no individual exposure assessment; the mere fact of having developed a particular disease deemed potentially service-connected and having had the *potential* for exposure is sufficient. The need for presumptions often arises because of a lack of exposure data (IOM 2008). In other situations, extensive "dose-reconstruction" investigations may be required for a person, such as an "atomic veteran," to qualify for compensation. For radiation-related exposures, Congress enacted the Radiation Exposure Compensation Act after a provision in the Orphan Drug Act that directed the National Institutes of Health to establish a set of "radio-epidemiologic tables" of estimates of probability of causation (PC), the probability that a given cancer was caused by a particular radiation dose (NIH 1985).

A key feature of that and many other compensation programs and toxic tort litigations is the uncertainty about a person's contact with the toxicant of concern that resulted in exposure. For radiation in particular, determining exposure has been facilitated by a Monte Carlo program from the National Institute for Occupational Safety and Health, the Integrated RadioEpidemiologic Program (NIOSH 2011), which provides a nuclear worker with not only a best estimate of his or her radiation dose but also an estimate of its uncertainty and carries this uncertainty through to the calculation of PC. Although the validity of the PC concept has been called into question (NRC 1984; Greenland and Robins 2000), there is no doubt that exposure information and its uncertainty—at either the individual or the population level—are central to establishing fair compensation policies. For example, IOM (2008) pointed out that policies aimed at erring on the side of avoiding false-negative decisions (failing to give compensation when it is due) in favor of false-positive decisions (excessively liberal policies that compensate people who do not deserve it), such as certain presumptive-disability provisions for veterans, would have the effect of rewarding ignorance—people with highly uncertain exposures (or with rare diseases for which the dose–response relationship is highly uncertain) are more likely to be compensated than those with more precise information. Such policies need further consideration to ensure an appropriate balance between fairness and societal burden. Overall, better assessment and characterization of worker exposures are needed to develop more accurate and protective exposure limits for the workplace. Better monitoring of workers can help to assure that current exposures are not harmful and improved exposure data can help to clarify compensation issues when people do become ill.

ENVIRONMENTAL PLANNING

Exposure assessment has contributed to urban and environmental planning, informing our understanding of how different patterns of land use can change the magnitudes of emissions and exposures of humans and ecosystems. Exposure science is increasingly used in health impact assessments.[2] Box 3-7 illustrates the application of exposure science to health impact assessment in San Francisco.

---

[2]A 2011 National Research Council report defined health impact assessment as a structured process that uses scientific data, professional expertise, and stakeholder input to identify and evaluate public-health consequences of proposals and suggests actions that could be taken to minimize adverse health impacts and optimize beneficial ones (NRC 2011).

BOX 3-7 Health Impact Assessment of Mobile Sources in San Francisco

Several health impact assessments conducted in 2004–2006 on land-use plans in Oakland and San Francisco identified conflicts between roadway proximity and sensitive land use. The conflicts resulted from placement of new high-density infill developments on land parcels along major highways (Bhatia 2007). Such conflicts were not addressed by federal, state, or local environmental standards, nor by local planning codes. The California Air Resources Board recommended not building housing within 500 ft of highways, but land-use planners typically did not follow this guidance.

In 2007, the San Francisco Department of Public Health (SFDPH), in its role as the environmental-health authority for the city, participated in an environmental review—required under the California Environmental Quality Act (CEQA)—of four neighborhood rezoning plans (Bhatia and Wernham 2008). The plans proposed large-scale industrial-to-residential rezoning in areas of the city near highways and busy arterial roads. SFDPH identified several important environmental impacts (air and noise pollution and traffic collisions) and proposed mitigations. To create a mitigation program compatible with the needs of local planners and CEQA regarding environmental impact assessment, SFDPH needed to identify an assessment method for each development project, a quantitative action level that triggered mitigation, and alternative engineering approaches for mitigation. To meet those needs, SFDPH proposed using a threshold concentration of fine particulate matter ($PM_{2.5}$) of 0.2 $\mu g/m^3$ from nearby roadway sources within 500 ft of a sensitive receptor as a proxy for important roadway hazards. The CALINE dispersion model (Benson 1989) was used to predict traffic-attributable $PM_{2.5}$, and a series of ventilation and filtration standards were proposed for new buildings within the 500-ft zone of roadways as a performance-based mitigation measure (Bhatia and Rivard 2008).

The approach was accepted by planners and the development community and adopted as mitigation for the four neighborhood plans (Bhatia and Wernham 2008). Later, SFDPH and the Department of Planning successfully sought the adoption of the approach as a public-health law applicable to all projects citywide. It remains the only local law in the United States that requires the assessment *and* mitigation of air pollution hazards from nearby roadways.

The current statute applies only to new residential development, but initiatives are underway to address existing development and a wider array of impacts, including noise pollution and dangers to pedestrians and bicyclists from traffic collisions. A key strategy for addressing hazards from nearby roadways is to develop metrics and modeling tools that can account for refined spatial characterizations of exposure and allow for assessments of attributable disease burden or other health-effect estimates (Wier et al. 2009a, 2009b; Bhatia and Seto 2011). The new law and associated policies rely heavily on information generated by exposure science and environmental epidemiology.

Urban Planning

Major economic and urban transformation over the last 30 years has led to a shift from large stationary emitters to larger concentrations of mobile sources and small manufacturing facilities, and more complex commuting and mobility patterns among workers. Those changes have accentuated small-area variations in pollution due to traffic and localized manufacturing. The restructuring has generally increased the amount of traffic and created spatial mismatches in residential and employment locations, lengthening commuting times and in-transit exposures of many workers (HEI 2010). As a consequence, environmental exposures affecting health have become more spatially heterogeneous and more dependent on human mobility and activities within and between places.

Exposures associated with urban transportation systems display variation over small areas that is difficult to quantify because of its spatial heterogeneity. Those exposures are pervasive in that they affect large populations. Recent studies by the Health Effects Institute (HEI 2010; Jerrett et al. 2011) have documented the proportion of the population that experiences high exposures from traffic, resulting in increased risks of accidents and exposures to air pollution and noise. In Toronto and Los Angeles, 43-45% of the population were in the "high-exposure zones" (that is, 500 meters from a major highway and 100 meters from a major road) (HEI 2010, p. 3-13). In the denser cities of Asia, 55-77% of the population were "highly exposed" (Jerrett et al. 2011, p. 35). Exposure science in urban areas faces the challenge of assessing spatially heterogeneous exposures that are pervasive throughout the urban structure. There is an increasing urgency to link exposure science to urban planning because of increasing population densities in cities, particularly through infill development, intended to reduce pressure from vehicle miles traveled to achieve climate change mitigation and other environmental goals. Often such "densification" efforts puts residents in greater proximity to transportation corridors, leading to increased exposures from traffic emissions and other related risks (for example, noise pollution) (Melia et al. 2011; Hankey et al. 2012). At the same time there will be continual stress on the water supply and the quality of tap water available.

Exposure science helps to inform urban planning, often providing insights into inherent trade-offs (Künzli et al. 2003). In many countries, for example, physical inactivity is a major threat to public health, and some of the inactivity is due to an urban environment that discourages active travel by foot or bicycle. However, active travel involves other exposures and potential health risks, including air pollution, accidents, and noise (de Nazelle et al. 2011). Exposure studies are being conducted to understand the effects of active travel. They require integration of ambient-pollutant concentration data with information on physical activity to assess potential inhalation and dose of air pollutants.[3]

---

[3]Information on ambient-pollutant concentrations combined with data on physical activity can inform potential dose, inasmuch as physically active people have inhalation

de Hartog et al. (2010) conducted an integrated health impact assessment that compared the likely effects of physical activity in bicycling with the risks posed by increased air-pollution exposure. The benefits of increased exercise far outweighed the increased risks from air pollution, but ideally urban planners can use exposure science to devise bicycling and walking routes that minimize the risks faced by active commuters, and exposure science can have a critical role in identifying the environments that are associated with the highest exposures (Marshall et al. 2010; de Nazelle et al. 2011).

Salutogenic (beneficial) exposures may generate health benefits. Many of the attributes (for example, parks) that promote human health also support ecosystem health and ecologic services. For example, urban forests contribute to carbon sequestration and protect water and air quality (Dwyer et al. 1992, McPherson et al. 1997) while encouraging behaviors positively associated with mental and physical health (Maas et al. 2006; Maller et al. 2006). Access to green space, parks, and recreational programming has been associated with increased physical activity or reduced obesity in large epidemiologic studies (Wolch et al. 2011). Studies in England have shown that in areas with greater access to green space, the socioeconomic group differences in cardiovascular risk is reduced (Mitchell and Popham 2008). In a study conducted in southern California, children equipped with GPSs and accelerometers were shown to have significant increases in physical activity while exposed to green space (Almanza et al. 2011). Recent studies also suggest that access to such resources may follow socioeconomic group differences, in that minority groups and socioeconomically disadvantaged groups have fewer options to access these resources (Dahmann et al. 2009). Also parks in poor and minority-group neighborhoods tend to be more polluted than those in neighborhoods of wealthier people and higher proportions of white people (Su et al. 2011). Exposure science can help urban planners to identify areas that lack salutogenic exposures that may buffer adverse exposures or produce direct health benefits. Salutogenic exposures also illustrate the link between healthy ecosystems and human health.

Ecosystem Planning

Although exposure assessments can include a diverse suite of ecologic receptors (for example, fish, aquatic plants, amphibians, birds), integrating ecologic information in a way that captures risk for the whole ecosystem has been difficult. In practice, a true ecosystem perspective may be restricted to the initial development of site conceptual models that identify potential exposure pathways, after which the focus may be on specific and limited receptors. Exposure science and risk assessment could benefit from adopting a holistic, systems-level perspective, similar to that embodied in ecosystem planning and management. Already, exposure assessments are beginning to integrate and apply information

---

rates that are 2–5 times greater than those of persons in public transit or in private vehicles, and that also may be higher because of the proximity to car traffic.

from the ecologic, socioeconomic, and political realms as does ecosystem planning (Grumbine 1994; Brussard et al. 1998; Szaro et al. 1998). Like ecosystem management (Grumbine 1994), exposure assessments also consider the hierarchic context of the process or problem, work collaboratively across ecologic boundaries, use diverse data from research and monitoring efforts, and, to some extent, recognize that humans are part of the ecosystem. However, one piece of ecosystem planning and management that is largely absent from current exposure assessments is the emphasis on adaptive management, whereby managers and planners treat management as a learning process (an experiment) and adjust actions as understanding improves. (Box 3-8 illustrates how exposure science has contributed to our understanding of the environmental impacts of stressors on Lake Tahoe.)

An ecosystem-based focus on land-use practices might be especially useful for exposure science. For example, regional changes in land-use practices can result in high levels of agricultural runoff and untreated sewage effluent that lead to eutrophication, in which aquatic ecosystems accumulate high concentrations of nutrients (such as phosphates and nitrates) that promote plant growth. Algal growth can become excessive and sometimes lead to harmful algal blooms (Paerl 1997; Cloern 2001; Anderson et al. 2002; Kemp et al. 2005). Likewise, land uses that reduce vegetative cover, especially trees, may result in changes in temperature that can directly prompt health-threatening exposures (for example, extreme heat events) or indirectly influence exposure to other substances. In aquatic ecosystems, degraded riparian zones, loss of forest cover, runoff from impervious surfaces, and discharges from industry can lead to rising water temperatures and increased toxicity. Above-normal temperatures compromise the function and integrity of aquatic ecosystems. They can also increase the sensitivity of aquatic animals to heavy metals and pesticides, including cadmium (Cherkasov et al. 2006, 2007), copper (Gupta et al. 1981, Khan et al. 2006), diazinon (Osterauer and Köhler 2008), and trichlorfon (Brecken-Folse et al. 1994).

An ecosystem approach to planning requires recognition that exposures may result from a series of diffuse and indirect interactions among humans and nonhuman species. Exposure to emerging infectious diseases, many of which originate in wildlife hosts or reservoirs, is an excellent illustration of this complexity. Human activities, particularly those related to habitat modification and resource subsidization, not only influence ecologic health but mediate wildlife–pathogen interactions through changes in host density, behavior, and spatial distribution (Dobson and Foufopoulos 2001; Gehrt 2010). Urban development, for instance, promotes high densities of some wildlife species (such as raccoon, *Procyon lotor*) that serve as hosts of many pathogens of public-health importance (such as rabies and leptospirosis: Bradley and Altizer 2006; Gehrt 2010) and can profoundly alter trophic interactions in ways that can result in loss of biodiversity (Faeth et al. 2005). Socioeconomic factors can exacerbate exposure if people lack the financial and educational resources that support preventive behaviors, such as vaccinating pets, excluding or removing wildlife from homes, and securing refuse and other potential wildlife attractants.

BOX 3-8 Exposure to Multiple Stressors in a Large Lake Ecosystem

Lake Tahoe is a large (19 × 35 km) and deep (501 m), ultraoligotrophic, montane lake in the Sierra Nevada of California and Nevada. It is known for its cobalt blue waters that are due, in part, to its small watershed area in comparison with its large lake volume. The economy of Lake Tahoe is intimately linked with its ecology—the economic value of the lake and surrounding basin is closely related to the perceived scenic and pristine character of the region's water, air, and forest resources (for example, Kearney et al. 2008). Recreation is the main industry of the region, with an economic impact from visitors of about $320 million per year, which supports nearly 10,000 jobs (Eiswerth et al. 2000; Dean Runyan Associates 2009). The intense pressure from local recreational activities, atmospheric deposition of nutrients and toxicants, intentional and unintentional introduction of nonnative species (for example, Eiswerth et al. 2000; Kamerath et al. 2008; Chandra et al. 2009), and climate change result in the simultaneous exposure of the ecosystem to multiple stressors (see Figure 3-2). Assessment of multiple stressor exposure in large, complex ecosystems is not easy, and the outcomes of the exposures are often unpredictable. However, Lake Tahoe presents a good case study of the ecologic and potential economic impacts of multiple stressor exposures.

Within the last 40 years, scientists have documented a steady decline in lake clarity and an increase in primary production (Jassby et al. 1996; Goldman 2000). That is especially the case with respect to ultraviolet (UV) radiation attenuation in the nearshore areas of the lake that are close to human developments and activities (Rose et al. 2009). Several factors play a role in the decrease in clarity, including the introduction of fine inorganic particles from the watershed; resuspension of fine particles due to wind, currents, and boats; and the settling of particles on the lake bottom (Jassby et al. 1996; Swift et al. 2006). Those changes have been accompanied by an increase in water temperature facilitated by increased absorption of solar radiation in turbid waters and due to climate change (Tucker et al. 2010). The combined and interactive effects of exposure to those stressors have had substantial adverse effects on the ecology of the watershed.

The role of interaction of water temperature, UV radiation, and resident nonnative species in the establishment of novel nonnative species, altered food-web structure, and increased exposure to toxic pollutants illustrates the effects of multiple stressor exposures. Nonnative crayfish became established in the lake in the 1930s (Chandra et al. 2009) and are widespread throughout the lake, Eurasian watermilfoil (an aquatic plant) was first discovered in the southern portion of the lake in the 1960s (Eiswerth et al. 2000), and warmwater centrachid fishes (largemouth bass and bluegill) were introduced by anglers in the middle to late 1990s (Kamerath et al. 2008; Chandra et al. 2009). By the middle 1990s, milfoil had spread to nearly all low-energy nearshore areas; by 2007, bass and bluegill were observed in marinas and nearshore areas containing milfoil and crayfish around the lake (Kamerath et al. 2008). Bass and bluegill are aggressive and can have predator-related and

(Continued)

BOX 3-8 Continued

competition-related effects on the native fishes in the nearshore areas (Gevertz et al. 2012). Their establishment has been possible only through the combined effects of increased water temperature, decreased UV radiation and water clarity, habitat alteration (milfoil), and abundant prey (crayfish and native fishes) (Kamerath et al. 2008; Tucker et al. 2010; Gevertz et al. 2012). Recently, water-quality thresholds for temperature, UV radiation, and water clarity required to prevent the establishment and spread of bass and bluegill have been suggested (Tucker et al. in press).

Ecologic and physiologic research has predicted that exposure to multiple stressors would facilitate the establishment of warm-water fish in Lake Tahoe, but an unpredicted outcome of the interactions may be an increase in exposure to toxic pollutants in the ecosystem. Mercury is an important toxic pollutant globally and in Lake Tahoe (Drevnick et al. 2010). Methylmercury (MeHg) biomagnifies in food webs, with the highest rates of transfer occurring in lower trophic levels (for example, water to algae) (Hammerschmidt and Fitzgerald 2006) and with top-predator fish and birds accumulating the most (Wiener et al. 2002). Crayfish can accumulate large amounts of MeHg as a benthic organism eating primarily periphyton and macrophytes (Momot et al. 1978) and largemouth bass commonly supplement their diets with crayfish (García-Berthou 2002). In addition to its primarily plant-based diet, crayfish are facultative omnivores and will feed on fish carcasses (Minckley and Craddock 1961). The biomagnification of persistent bioaccumulative toxicants, such as MeHg, may be increased by a trophic feedback cycle in which bass eat crayfish, which eat bass, and so on. Laboratory experiments and modeling demonstrate that such a trophic feedback cycle substantially increases MeHg accumulation in bass and crayfish (Bowling et al. 2011). The results are supported by anecdotal evidence that another, native species (brown trout) has higher tissue concentrations of MeHg in lakes that have crayfish than in lakes that do not (Oris et al. 2004). Thus, it is possible that exposure to multiple stressors that have facilitated the establishment of largemouth bass in Lake Tahoe can result in large increases in MeHg exposure of both native and nonnative species and affect the ecologic resources of the region. These native and non-native species are also direct vectors of MeHg exposure of humans and could decrease the recreational value of the lake.

Changes in species composition due to human activities may affect exposure through alterations in species interactions. For example, human-facilitated invasion of Amur honeysuckle (*Lonicera maackii*), a nonnative shrub, increases human risk of ehrlichiosis, an emerging infectious disease transmitted by ticks (Allan et al. 2010). The high risk results from a preference of a key tick and pathogen reservoir, white-tailed deer (*Odocoileus virginianus*), for areas with dense honeysuckle. In aquatic systems, the invasive round goby (*Neogobius melanostomus*) is thought to facilitate mobilization of contaminants in food webs and to increase exposure of humans because its persistence in contaminated environments draws predatory fish, which also are popular game species, into polluted habitats (Marentette et al. 2010). Thus, understanding human exposure to pathogens requires a holistic ecosystem perspective.

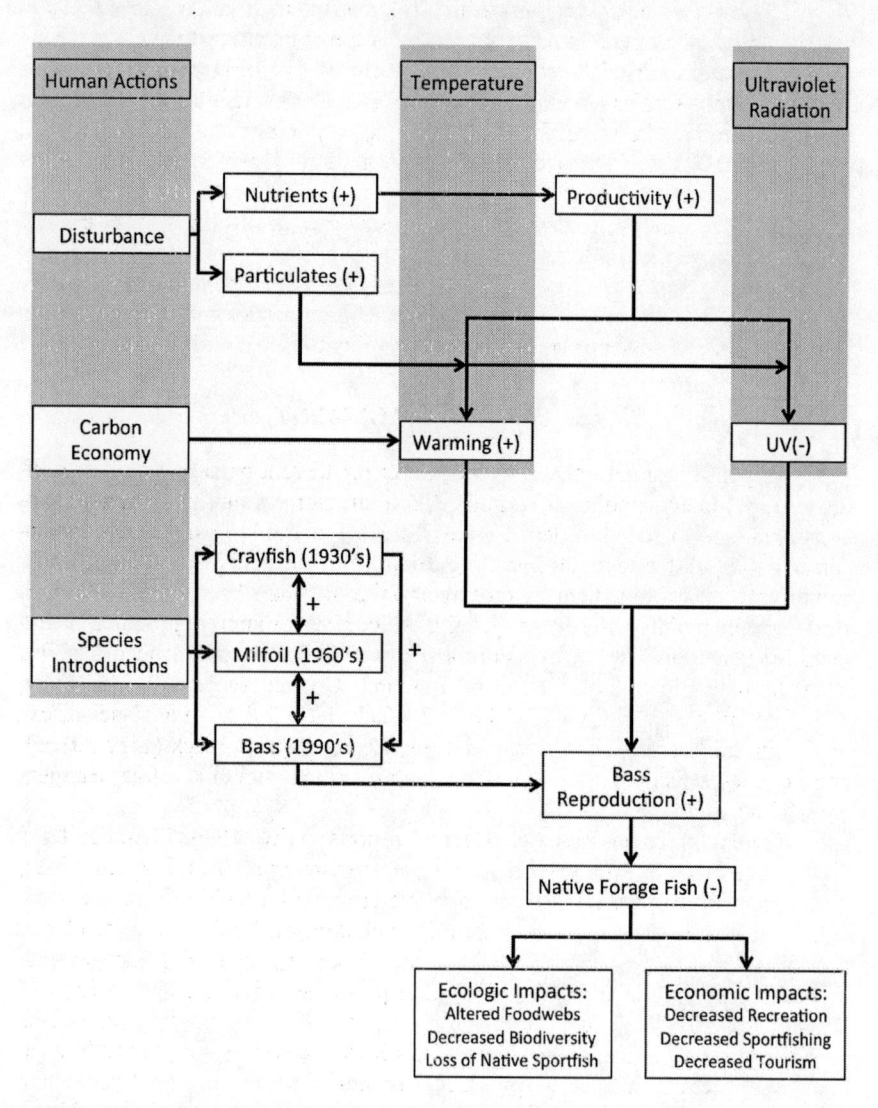

FIGURE 3-2 Exposure to Multiple Stressors in Lake Tahoe.

An ecosystem approach extends beyond "natural" systems to include the built environment. Cities are coupled human–nature systems that can be understood only through their linked ecologic, physical, and socioeconomic components (Pickett et al. 2001; Alberti et al. 2003). Spatially explicit approaches common to ecosystem planning fit urban ecosystems well in that they can be viewed as mosaics of risk and protection (Fitzpatrick and LaGory 2011). Characterizing the heterogeneity of social and environmental attributes in urban ecosystems better is especially useful in a climate of diminishing resources. For example, recent studies demonstrate that bioavailability of contaminants in soils is strongly mediated by environmental heterogeneity (Filser et al. 2008), and this environmental heterogeneity can be characterized with sophisticated mapping techniques to understand exposure risk (Lahr and Kooistra 2010). By understanding heterogeneity, planners and managers can identify the most critical areas for remediation, saving resources and sometimes avoiding ecologically damaging consequences of unnecessary efforts (such as removing topsoil from large areas). Understanding complex socio-ecologic processes that are operating in urban ecosystems is a prerequisite for protecting ecologic and human health.

DISASTER MANAGEMENT

Response to a natural, accidental, or terrorist event or to an act of war often requires understanding of exposure to a stressor or multiple stressors. Instrumentation is needed to detect biologic, chemical, and physical stressors accurately and cost effectively, and, depending on the evolving situation, both community and occupational exposure monitoring may be required. Defined strategies for rapidly assessing exposures are critical for applying predictive and measurement tools. Active exposure assessment is an important need that has been discussed in previous National Research Council reports in connection with the deployed military (NRC 1999; 2000a,b; IOM 2000), in analyses of exposure in the aftermath of the World Trade Center (WTC) attack (Lioy 2010a), and in general disaster discussions and analyses (Dominici et al. 2005; Bongers et al. 2008; Rodes et al. 2008).

Several National Research Council reports (NRC 1999; 2000a,b; IOM 2000) have discussed the need for exposure assessment after military activities, during combat, or where troops are deployed to guard a facility or area associated with identified or suspected hazards, including oil fields, chemical plants, contraband destruction (incinerator) sites, and power plants. Exposure assessments are intended as a means of reducing personnel injuries and as a form of preventive surveillance as information on exposures can be used by physicians at a later time to treat exposed individuals if they develop symptoms. In such situations, there is a need to provide information on the personal protective equipment required and to enforce administrative controls to guard sites. Monitoring needs to be flexible and surveys need to be conducted to characterize the exposures that may be encountered before establishing permanent security or conducting similar activities. Biologic monitoring and surveillance may need to be formalized for deployed troops.

In the aftermath of the WTC disaster, the nature of the exposures and the populations exposed changed at Ground Zero (Lioy and Gochfeld 2002; EPA 2002, 2003; GAO 2007; Lorber et al. 2007). The evolving situation resulted in exposure of responders and others under different conditions. Box 3-9 discusses issues surrounding emergency management after the attack on the WTC.

It is important to develop and test a series of protocols to ensure that community and occupational monitoring and analytic capabilities are field-ready in the event of an emergency situation. Exposure-monitoring platforms may need to include personal sensors and portable monitors mounted on vehicles and be portable and robust enough to be airlifted to remote sites. A variety of samplers can be tested, including ones that measure physical, chemical, and biologic materials. Surveillance of workers and the community with biologic monitoring is essential for understanding continuing exposures or the cessation of exposures at least through rescue, re-entry, and recovery. Similarly, a field laboratory to process biomonitoring samples would be important. Strategies to minimize continuing community exposure are essential, for example, the use of exclusion zones for various periods. Effective personal protective devices can be made available to those in harm's way to minimize injury or disease, and leaders could define and enforce the use of this equipment.

During disasters, there need to be clear and effective lines of communication, members of a team that can interpret data, and officers who understand the meaning of the exposures that are determined. Most important, guidance and recommendations from professionals on how to proceed are needed.

CONCLUSIONS

Exposure science has made and continues to make contributions to many fields. Increased emphasis on integration and new applications of the principles of exposure science can contribute major benefits to the areas of epidemiology, toxicology, environmental regulation, ecologic planning, and disaster management.

Epidemiology

Exposure science should aim to provide epidemiologists with novel tools to improve exposure assessment, particularly comprehensive assessment of cumulative exposures to the ensemble of relevant stressors. That will require use of new remote-sensing and other high-volume techniques for measuring the external environment and new –omics technologies for the internal environment. Population-wide exposure databases (such as a national dose registry for radiation exposures) that could be routinely linked with population-wide outcome databases (such as cancer registries or the National Death Index) would facilitate rapid discovery of novel hazards.

BOX 3-9 Emergency Management After the Attack on the World Trade Center

The collapse of the WTC towers was a horrific aftermath of the 9/11 terrorist attacks. The generation of dust due to the disintegration of building materials and the ensuing fires created demands for exposure information to understand the associated potential health effects on workers and the community—in the immediate aftermath of the disaster, in the days immediately afterwards when the dust was resuspended, and during cleanup in the weeks and months after the disaster. The need for such data was not anticipated before the event (Lioy and Gochfeld 2002; Lioy 2010a).

The WTC collapse illustrates the need for an integrated and timely approach to data collection involving complex mixtures (EPA 2003). Data on the presence of radiation and other hazardous agents were initially collected by several agencies but with little coordination. Inventories of potential sources of known hazardous chemicals (such as pesticides) were documented. Biologic samples were collected from firefighters to examine the concentrations of gases and other contaminants that may have been inhaled at Ground Zero (Edelman et al. 2003). However, the composition of WTC dust remained unknown, and this crippled the initial assessments of the potential hazards (Lioy 2010a). The failure to characterize exposures adequately stemmed from a lack of real-time portable devices to measure particles and gaseous materials in dusty environments and from a lack of understanding that the immediate concerns were short-term exposures (Lioy 2010a). The mindset at the time was focused on low-level exposures (for example, to asbestos and fine particles) associated with long-term disease outcomes.

The exposure-assessment issues stemming from the WTC disaster led to discussions of how to characterize exposure during rescue, reentry, recovery, restoration, and rehabilitation after an event (Rodes et al. 2008; Lioy 2010b). An important aspect of the WTC disaster was the recognition that the combination of collapse of the buildings with intense fires resulted in emissions of complex mixtures because of the nature of the construction materials, furnishings, and equipment (EPA 2002, 2003; GAO 2007). Training of emergency and law-enforcement agency personnel regarding potential exposures and potential hazards to health is important, including rapid exposure characterization to minimize additional injuries to workers and the public.

Risk assessment, risk management, and risk communication are valuable evaluation tools, but their effectiveness depends on the accuracy of information and on the speed with which it can be interpreted and conveyed. A lesson for exposure science from the disaster is that it is important to assume a worst-case scenario and efforts would need to be made in advance to implement exposure-characterization strategies, including prepositioning of appropriate measuring devices. The application of exposure-science principles during acute events can augment activities associated with a national response plan.

Toxicology

Exposure science needs to interface with toxicology to inform selection of and priority-setting among chemicals at environmentally relevant concentrations for toxicity testing, to develop human pharmacokinetic data, and to improve understanding of the exposure–response relationships involved in perturbations of toxicity pathways in exposed humans or wildlife. Closer interactions between toxicologists and epidemiologists to characterize the biologic effectiveness of exposures could provide better exposure metrics.

Environmental Regulation

Exposure science needs to play a greater role in the process of risk assessment, risk management, and compensation policy to improve characterization of population-wide exposure distributions, their uncertainties, aggregate and cumulative exposures, and high-risk populations. More powerful incentives for data generation—such as conservative default values—are needed to address critical data gaps. A more risk-preventive approach to introduction of new chemicals into the environment and better post-marketing surveillance of those already present is needed, particularly with the goal of reducing disparities in the population. An improved role of exposure science in the regulations will advance priority-setting, help measure the efficacy of regulations, and reduce the likelihood of widespread human or environmental exposure to substances that are likely to be hazardous.

Environmental Planning

Exposure science should be an increasingly integral component of environmental planning because of its ability to inform decisions during urban projects that may affect public health (for example, through exposure to increased air-pollutant emissions or potential increases in exposure to green space) and because of its ability to recognize the complex interactions among humans and ecosystems that are critical for protecting human and ecosystem health.

Disaster Management

A great challenge in the management of natural or human-made disasters is to anticipate the needs for exposure science before a disaster occurs. To that end, field-ready protocols need to be established and tested, including the necessary equipment (such as drones) and expert personnel, and the equipment needs to be pre-positioned for rapid deployment to conduct community and occupational monitoring. The availability of personal protective devices is an important element in these preparations. Biologic monitoring for surveillance of workers and the community is essential. Ideally, it would be conducted routinely on first-responder personnel, in advance of a disaster, to measure toxicants associated with their jobs and thus provide a baseline measure. However, biomonitoring cannot take the place of environmental sensors for real-time measurement of hazardous agents.

REFERENCES

Alberti, M., J. M. Marzluff, E. Shulenberger, G. Bradley, C. Ryan, and C. Zumbrunnen. 2003. Integrating humans into ecology: Opportunities and challenges for studying urban ecosystems. Bioscience 53(12):1169-1179.

Allan, B.F., H.P. Dutra, L.S. Goessling, K. Barnett, J.M. Chase, R.J. Marquis, G. Pang, G.A. Storch, R.E. Thach, and J.L. Orrock. 2010. Invasive honeysuckle eradication reduces tick-borne disease risk by altering host dynamics. Proc. Natl. Acad. Sci. USA 107(43):18523-18527.

Almanza, E., M. Jerrett, G. Dunton, E. Seto, and M. Pentz. 2011. Green Spaces in Healthy Places: Objective Data Demonstrates an Association between Greenness and Momentary Measures of Physical Activity in Children. Presentation at Active Living Research (ALR) Conference, February 22-24 2011, San Diego, CA [online]. Available: http://www.activelivingresearch.org/files/2011_GPS_Almanza.pdf [accessed Dec. 15, 2011].

Andersen, M.E., H.J. Clewell, III, E. Bermudez, D.E. Dodd, G.A. Wilson, J.L. Campbell, and R.S. Thomas. 2010. Formaldehyde: Integrating dosimetry, cytotoxicity, and genomics to understand dose-dependent transitions for an endogenous compound. Toxicol. Sci. 118(2):716-731.

Anderson, D.M., P.M. Glibert, and J.M. Burkholder. 2002. Harmful algal blooms and eutrophication: Nutrient sources, composition, and consequences. Estuaries 25(4b):704-726.

Barr, D.B., R.Y. Wang, and L.L. Needham. 2005. Biologic monitoring of exposure to environmental chemicals throughout the life stages: Requirements and issues for consideration for the National Children's Study. Environ. Health Perspect. 13(8):1083-1091.

Bell, M.L., K. Ebisu, and R.D. Peng. 2011. Community-level spatial heterogeneity of chemical constituent levels of fine particulates and implications for epidemiological research. J. Expo. Sci. Environ. Epidemiol. 21(4):372-384.

Benson, P. 1989. Caline4 - A Dispersion Model for Predicting Air Pollution Concentration Near Roadways. Report No. FHWA/CA/TL-84/15. California Department of Transportation, Sacramento, CA [online]. Available: http://www.weblakes.com/products/calroads/resources/docs/CALINE4.pdf [accessed May 9, 2012].

Bhatia, R. 2007. Protecting health using an environmental impact assessment: A case study of San Francisco land use decision-making. Am. J. Public Health 97(3):406-413.

Bhatia, R., and T. Rivard. 2008. Assessment and Mitigation of Air Pollutant Health Effects from Intra-urban Roadways: Guidance for Land Use Planning and Environmental Review. San Francisco, CA: San Francisco Department of Public Health [online]. Available: http://www.sfphes.org/publications/Mitigating_Roadway_AQLU_Conflicts.pdf [accessed Dec. 14, 2011].

Bhatia, R., and E. Seto. 2011. Quantitative estimation in health impact assessment: Opportunities and challenges. Environ. Impact Assess. Rev. 31(3):301-309.

Bhatia, R., and A. Wernham. 2008. Integrating human health into environmental impact assessment: An unrealized opportunity for environmental health and justice. Environ. Health Perspect. 116(8):991-1000.

Bongers, S., N.A.H. Janssen, B. Reiss, L. Grievink, E. Lebret, and H. Kromhout. 2008. Challenges of exposure assessment for health studies in the aftermath of chemical incidents and disasters. J. Expo. Sci. Environ. Epidemiol. 18(4):341-359.

Bowling, A.M., C.R. Hammerschmidt, and J.T. Oris. 2011. Necrophagy by a benthic omnivore influences biomagnification of methylmercury in fish. Aquat. Toxicol. 102(3-4):134-141.

Bradley, C.A., and S. Altizer. 2006. Urbanization and the ecology of wildlife diseases. Trends Ecol. Evol. 22(2):95-102.

Bradman, A., and R.M. Whyatt. 2005. Characterizing exposures to nonpersistent pesticides during pregnancy and early childhood in the National Children's Study: A review of monitoring and measurement methodologies. Environ. Health Perspect. 113(8):1092-1099.

Brecken-Folse, J.A., F.L. Mayer, L.E. Pedigo, and L.L. Marking. 1994. Acute toxicity of 4-nitrophenol, 2,4-dinitrophenol, terbufos and trichlorfon to grass shrimp (*Palaemonetes Spp*) and sheepshead minnows (*Cyprinodon variegatus*) as affected by salinity and temperature. Environ. Toxicol. Chem. 13(1):67-77.

Breslow, N.E., T. Lumley, C.M. Ballantyne, L.E. Chambless, and M. Kulich. 2009. Using the whole cohort in the analysis of case-cohort data. Am. J. Epidemiol. 169(11):1398-1405.

Brown, P., J.G. Brody, R. Morello-Frosch, J. Tovar, A.R. Zota, and R.A. Rudel. 2012. Measuring the success of community science: The Northern California Household Exposure Study. Environ. Health Perspect. 120(3):326-331.

Brussard, P.F, M.J. Reed, and T.C. Richard. 1998. Ecosystem management: What is it really? Landscape Urban Plan. 40:9-20.

Carroll, R.J., D. Ruppert, L.A. Stefanski, and C. Crainiceanu. 2006. Measurement Error in Nonlinear Models: A Modern Perspective, 2nd Ed. Boca Raton, FL: Chapman & Hall/CRC.

CDC (Centers for Disease Control and Prevention). 2011. Fourth National Report on Human Exposure to Environmental Chemicals. U.S. Department of Health and Human Services, Centers for Disease Control and Prevention, Atlanta, GA [online]. Available: http://www.cdc.gov/exposurereport/ [accessed Dec. 12, 2011].

CDPH (California Department of Public Health). 2010a. Public Drinking Water System [online]. Available: http://www.cdph.ca.gov/certlic/drinkingwater/Pages/default.aspx [accessed Jan. 3, 2012].

CDPH (California Department of Public Health). 2010b. Implementation of the California Environmental Contaminant Biomonitoring Program. Report to the California Legislature. California Department of Public Health in Collaboration with California Environmental Protection Agency's Office of Health Hazard Assessment and Department of Toxic Substances Control, January 2010 [online]. Available: http://oehha.ca.gov/multimedia/biomon/pdf/CECBPLegReport.pdf [accessed Dec. 6, 2011].

Chandra, S., H. Segale, and S. Adler. 2009. Lake Tahoe Species Introduction Timeline. Tahoe Environmental Research Center [online]. Available: http://terc.ucdavis.

edu/education_outreach/educationprograms/TIC_InvasiveSpeciesTimeline.pdf [accessed 17 May, 2012].

Cherkasov, A.S., P.K. Biswas, D.M. Ridings, A.H. Ringwood, and I.M. Sokolova. 2006. Effects of acclimation temperature and cadmium exposure on cellular energy budgets in the marine mollusk *Crassostrea virginica*: Linking cellular and mitochondrial responses. J. Exp. Biol. 209(7):1274-1284.

Cherkasov, A.S., S. Grewal, and I.M. Sokolova. 2007. Combined effects of temperature and cadmium exposure on haemocyte apoptosis and cadmium accumulation in the eastern oyster *Crassostrea virginica* (Gmelin). J. Therm. Biol. 32(3):162-170.

Chow, J.C., J.P. Engelbrecht, N.C. Freeman, J.H. Hashim, M. Jantunen, J.P. Michaud, S. Saenz de Tejada, J.G. Watson, F. Wei, W.E. Wilson, M. Yasuno, and T. Zhu. 2002. Chapter one: Exposure measurements. Chemosphere 49(9):873-901.

Cloern, J.E. 2001. Our evolving conceptual model of the coastal eutrophication problem. Mar. Ecol. Prog. Ser. 210:223-253.

Cohen Hubal, E.A., A.M. Richard, I. Shah, J. Gallagher, R. Kavlock, J. Blancato, and S.W. Edwards. 2010. Exposure science and the U.S. EPA National Center for Computational Toxicology. J. Expo. Sci. Environ. Epidemiol. 20(3):231-236.

Cohen Hubal, E.A., D.B. Barr, H.M. Koch, and T. Bahadori. 2011. The promise of exposure science. J. Expo. Sci. Environ. Epidemiol. 21(2):121-122.

Dahmann, N., J. Wolch, P. Joassart-Marcelli, K. Reynolds, and M. Jerrett. 2009. The active city? Disparities in provision of urban public recreation resources. Health Place 16(3):431-445.

Dale, V., G.R. Biddinger, M.C. Newman, J.T. Oris, G.W. Suter, T. Thompson, T.M. Armitage, J.L. Meyer, R.M. Allen-King, G.A. Burton, P.M. Chapman, L.L. Conquest, I.J. Fernandez, W.G. Landis, L.L. Master, W.J. Mitsch, T.C. Mueller, C.F. Rabeni, A.D. Rodewald, J.G. Sanders, and I.L. van Heerden. 2008. Enhancing the ecological risk assessment process. Integr. Environ. Assess. Manag. 4(3):306-313.

Dean Runyan Associates. 2009. The Economic Significance of Travel the North Lake Tahoe Area: 2003-2008 Detailed Visitor Impact Estimates. Prepared for North Lake Tahoe Resort Association, Tahoe City, CA, by Dean Runyan Associates, Portland, OR [online]. Available: http://www.deanrunyan.com/doc_library/FinalReportCA.pdf [accessed May 21, 2012].

de Hartog, J.J., H. Boogaard, H. Nijland, and G. Hoek. 2010. Do the health benefits of cycling outweigh the risks? Environ. Health Perspect. 118(8):1109-1116.

de Nazelle, A., M.J. Nieuwenhuijsen, J.M. Anto, M. Brauer, D. Briggs, C. Braun-Fahrlander, N. Cavill, A.R. Cooper, H. Desqueyroux, S. Fruin, G. Hoek, L.I. Panis, N. Janssen, M. Jerrett, M. Joffe, Z. Jovanovic Andersen, E. van Kempen, S. Kingham, N. Kubesch;, K.M. Leyden, J.D. Marshall, J. Matamala, G. Mellios, M. Mendez, H. Nassif, D. Oglivie, R. Peiro, K. Perez, A. Rabl, M. Ragettli, D. Rodriguez, D. Rojas, P. Ruiz, J.F. Sallis, J. Terwoert, J.F. Toussaint, J. Tuomisto, M. Zuurbier, and E. Lebret. 2011. Improving health through policies that promote active travel: A review of evidence to support integrated health impact assessment. Environ. Int. 37(4):766-777.

Dix, D.J., K.A. Houck, M.T. Martin, A.M. Richard, R.W. Setzer, and R.J. Kavlock. 2007. The ToxCast program for prioritizing toxicity testing of environmental chemicals. Toxicol. Sci. 95(1):5-12.

Dobson, A., and J. Foufopoulos. 2001. Emerging infectious pathogens of wildlife. Philos. Trans. R. Soc. London B 356(1411):1001-1012.

Dominici, F., J.I. Levy, and T.A. Louis. 2005. Methodological challenges and contributions in disaster epidemiology. Epidemiol. Rev. 27(1):9-12.

Downs, T.J., Y. Ogneva-Himmelberger, O. Aupont, Y. Wang, A. Raj, P. Zimmerman, R. Goble, O. Taylor, L. Churchill, C. Lemay, T. McLaughlin, and M. Felice. 2010. Vulnerability-based spatial sampling stratification for the National Children's Study, Worcester County, Massachusetts: Capturing health-relevant environmental and sociodemographic variability. Environ. Health Perspect. 118(9):1318-1325.

Drevnick, P.E., A. Shinneman, C. Lamborg, D.R. Engstrom, M. Bothner, and J.T. Oris. 2010. Mercury flux to sediments of Lake Tahoe, California-Nevada. Water Air Soil Pollut. 210 (1-4):399-407.

Dwyer, J.F., E.G. McPherson, H.W. Schroeder, and A. Rowntree. 1992. Assessing the benefits and costs of the urban forest. J. Arborculture 18(5):227-234.

Edelman, P., J. Osterloh, J. Pirkle, S.P. Caudill, J. Grainger, R. Jones, B. Blount, A. Calafat, W. Turner, D. Feldman, S. Baron, B. Bernard, D.B. Lushniak, K. Kelly, and D. Prezant. 2003. Biomonitoring of chemical exposure among New York City firefighters responding to the World Trade Center fire and collapse. Environ. Health Perspect. 111(16):1906-1911.

Eiswerth, M.E., S.G. Donaldson, and W.S. Johnson. 2000. Potential environmental impacts and economic damages of Eurasian watermilfoil (Myriophyllum spicatum) in western Nevada and northeastern California. Weed Technol. 14(3):511-518.

EPA (U.S. Environmental Protection Agency). 1992. Guidelines for Exposure Assessment. EPA600Z-92/001. Risk Assessment Forum, U.S. Environmental Protection Agency, Washington, DC.

EPA (U.S. Environmental Protection Agency). 2000. Guidance for Assessing Chemical Contaminant Data for Use in Fish Advisories. Volume 2: Risk Assessment and Fish Consumption Limits -Third Edition. EPA 823-B-00-008. Office of Science and Technology, Office of Water, U.S. Environmental Protection Agency, Washington, DC [online]. Available: http://water.epa.gov/scitech/swguidance/fishshellfish/techguidance/risk/upload/2009_04_23_fish_advice_volume2_v2cover.pdf [accessed May 9, 2012].

EPA (U.S. Environmental Protection Agency). 2002. Exposure and Human Health Evaluation of Airborne Pollution from the World Trade Center Disaster. External Review Draft. EPA/600/P-2/002A. National Center for Environmental Assessment, Office of Research and Development, U.S. Environmental Protection Agency, Washington, DC [online]. Available: http://nycosh.org/uploads/listit/id21/WTC_FINAL5.pdf [accessed Jan. 4, 2012].

EPA (U.S. Environmental Protection Agency). 2003. EPA's Response to the World Trade Center Collapse: Challenges, Successes, and Areas for Improvement. Report No. 2003-P-00012. Office of Inspector General, U.S. Environmental Protection Agency, Washington, DC. August 21, 2003 [online]. Available: http://www.epa.gov/oig/reports/2003/WTC_report_20030821.pdf [accessed Dec. 12, 2011].

EPASAB (U.S. Environmental Protection Agency Science Advisory Board). 2007. Advice to EPA on Advancing the Science and Application of Ecological Risk Assessment in Environmental Decision Making. EPA-SAB-08-002. Science Advisory Board, U.S. Environmental Protection Agency, Washington, DC [online]. Available: http://yosemite.epa.gov/sab/sabproduct.nsf/7140DC0E56EB148A8525737900043063/$File/sab-08-002.pdf [accessed Dec. 14, 2011].

Faeth, S.H., P.S. Warren, E. Shochat, and W.A. Marussich. 2005. Trophic dynamics in urban communities. Bioscience 55(5):399-407.

Filser, J., H. Koehler, A. Ruf, J. Rombke, A. Prinzing, and M. Schaefer. 2008. Ecological theory meets soil ecotoxicology: Challenge and chance. Basic Appl. Ecol. 9(4):346-355.

Fitzpatrick, K., and M. LaGory. 2011. Unhealthy Cities: Poverty, Race, and Place in America, 2nd Ed. New York: Routledge.

GAO (U.S. Government Accountability Office). 2007. World Trade Center: EPA's Most Recent Test and Clean Program Raises Concerns That Need to Be Addressed to Better Prepare for Indoor Contamination Following Disasters. GAO 07-1091. Washington, DC: U.S. Government Accountability Office. September 2007 [online]. Available: http://www.gao.gov/new.items/d071091.pdf [accessed Dec. 12, 2011].

GAO (General Accountability Office). 2009. Chemical Regulation: Options for Enhancing the Effectiveness of the Toxic Substances Control Act. GAO-09-428T. Washington, DC: U.S. Government Accountability Office [online]. Available: http://www.gao.gov/new.items/d09428t.pdf [accessed May 9, 2011].

García-Berthou, E. 2002. Ontogenetic diet shifts and interrupted piscivory in introduced largemouth bass (*Micropterus salmoides*). Int. Rev. Hydrobiol. 87(4):353-363.

Gehrt, S.D. 2010. The urban ecosystem. Pp. 3-11 in Urban Carnivores: Ecology, Conflict, and Conservation, S.D. Gehrt, S.P.D. Riley, and B.L. Cypher, eds. Baltimore, MD: The Johns Hopkins University Press.

Georgopoulos, P.G., A.F. Sasso, S.S. Isukapalli, P.J. Lioy, D.A. Vallero, M. Okino, and L. Reiter. 2009. Reconstructing population exposures to environmental chemicals from biomarkers: Challenges and opportunities. J. Expo. Sci. Environ. Epidemiol. 19(2):149-171.

Gevertz, A.K., A.J. Tucker, A.M. Bowling, C.E. Williamson, and J.T. Oris. 2012. Differential tolerance of native and non-native fish exposed to ultraviolet radiation and fluoranthene in Lake Tahoe (CA/NV). Environ. Toxicol. Chem. 31(5): 1129-1135.

Gilliland, F.D., Y.F. Li, A. Saxon, and D. Diaz-Sanchez. 2004. Effect of glutathione-S-transferase M1 and P1 genotypes on xenobiotic enhancement of allergic responses: Randomised, placebo-controlled crossover study. Lancet 363(9403): 119-125.

Gilliland, F., E. Avol, P. Kinney, M. Jerrett, T. Dvonch, F. Lurmann, T. Buckley, P. Breysse, G. Keeler, T. de Villiers, and R. McConnell. 2005. Air pollution exposure assessment for epidemiologic studies of pregnant women and children: Lessons learned from the Centers for Children's Environmental Health and Disease Prevention Research. Environ. Health Perspect. 113(10):1447-1454.

Goldman, C.R. 2000. Four decades of change in two subalpine lakes. Verh. Internat. Verein. Limnol. 27(Pt. 1):7-26.

Greenland, S., and J.M. Robins. 2000. Epidemiology, justice, and the probability of causation. Jurimetrics 40:321-340.

Grumbine, R.E. 1994. What is ecosystem management? Conserv. Biol. 8(1):27-38.

Gupta, P.K., B.S. Khangarot, and V.S. Durve. 1981. The temperature-dependence of the acute toxicity of copper to a fresh-water pond snail, *Viviparus bengalensis* L. Hydrobiologia 83(3):461-464.

Hammerschmidt, C., and W. Fitzgerald. 2006. Bioaccumulation and trophic transfer of methylmercury in the Long Island Sound. Arch. Environ. Contam. Toxicol. 51(3):416-424.

Hankey, S., J.D. Marshall, and M. Brauer. 2012. Health impacts of the built environment: Within-urban variability in physical inactivity, air pollution, and ischemic heart disease mortality. Environ. Health Perspect. 120(2):247-253.

HEI (Health Effects Institute). 2010. Traffic-Related Air Pollution: A Critical Review of the Literature on Emissions, Exposure, and Health Effects. Special Report 17. Boston, MA: Health Effects Institute [online]. Available: http://pubs.healtheffects.org/view.php?id=334 [accessed Dec. 12, 2011].

Hirschfeld, S., D. Songco, B.S. Kramer, and A.E. Guttmacher. 2011. National Children's Study: Update in 2010. Mt Sinai J. Med. 78(1):119-125.

Hoffman, F.O., A.J. Ruttenber, A.I. Apostoaei, R.J. Carroll, and S. Greenland. 2007. The Hanford Thyroid Disease Study: An alternative view of the findings. Health Phys. 92(2):99-111.

IOM (Institute of Medicine). 2000. Protecting Those Who Serve: Strategies to Protect the Health of Deployed U.S. Forces. Washington, DC: National Academy Press.

IOM (Institute of Medicine). 2008. Improving the Presumptive Disability Decision-Making Process for Veterans, J.M. Samet, and C.C. Bodurow, eds. Washington, DC: National Academies Press.

Jassby, A.D., J.E. Reuter, and C.R. Goldman. 1996. Determining long-term water quality in the presence of climate variability: Lake Tahoe (USA). Can. J. Fish. Sci. 60(12):1452-1461.

Jerrett, M., R.T. Burnett, R. Ma, C.A. Pope, D. Krewski, K.B. Newbold, G. Thurston, Y. Shi, N. Finkelstein, E.E. Calle, and M.J. Thun. 2005. Spatial analysis of air pollution and mortality in Los Angeles. Epidemiology 16(6):727-736.

Jerrett, M., J. Su, J. Apte, and B. Beckerman. 2011. Estimates of population exposure to traffic-related air pollution in Beijing, China and New Delhi, India: Extending exposure analyses report in HEI Special Report 17, Traffic-Related Air Pollution: Critical Review of the Literature on Emissions, Exposure, and Health Effects, November 5, 2010. Boston, MA: Health Effects Institute [online]. Available: http://www.healtheffects.org/International/Jerrett_Asia_Traffic_Exposure.pdf [accessed Dec. 10, 2011].

Kamerath, M., S. Chandra, and B.C. Allen. 2008. Distribution and impacts of war water invasive fish in Lake Tahoe, USA. Aquat. Invasions 3(1):35-41.

Kearney, A.R., G.A. Bradley, C.H. Petrich, R. Kaplan, S. Kaplan, and D. Simpson-Colbank. 2008. Public perception as support for scenic quality regulation in a nationally treasured landscape. Landscape Urban Plan. 87(2):117-128.

Kemp, W.M., W.R. Boynton, J.E. Adolf, D.F. Boesch, W.C. Boicourt, G. Brush, J.C. Cornwell, T.R. Fisher, P.M. Glibert, J.D. Hagy, L.W. Harding, E.D. Houde, D.G. Kimmel, W.D. Miller, R.I.E. Newell, M.R. Roman, E.M. Smith, and J.C. Stevenson. 2005. Eutrophication of Chesapeake Bay: Historical trends and ecological interactions. Mar. Ecol. Prog. Ser. 303:1-29.

Khan, M.A., S.A. Ahmed, B. Catalin, A. Khodadourst, O. Ajayi, and M. Vaughn. 2006. Effect of temperature on heavy metal toxicity to juvenile crayfish, *Orconectes immunis* (Hagen). Environ. Toxicol. 21(5):513-520.

Kopecky, K.J., S. Davis, T.E. Hamilton, M.S. Saporito, and L.E. Onstad. 2004. Estimation of thyroid radiation doses for the Hanford thyroid disease study: Results and implications for statistical power of the epidemiological analyses. Health Phys. 87(1):15-32.

Kopecky, K.J., L. Onstad, T.E. Hamilton, and S. Davis. 2005. Thyroid ultrasound abnormalities in persons exposed during childhood to 131I from the Hanford nuclear site. Thyroid 15(6):604-613.

Künzli, N., R. McConnell, D. Bates, T. Bastain, A. Hricko, F. Lurmann, E. Avol, F. Gilliland, and J. Peters. 2003. Breathless in Los Angeles: The exhausting search for clean air. Am. J. Public Health 93(9):1494-1499.

Lahr, J., and L. Kooistra. 2010. Environmental risk mapping of pollutants: State of the art and communication aspects. Sci. Total Environ. 408(18):3899-3907.

Landrigan, P.J., L. Trasande, L.E. Thorpe, C. Gwynn, P.J. Lioy, M.E. D'Alton, H.S. Lipkind, J. Swanson, P.D. Wadhwa, E.B. Clark, V.A. Rauh, F.P. Perera, and E. Susser. 2006. The National Children's Study: A 21-year prospective study of 100,000 American children. Pediatrics 118(5):2173-2186.

Lioy, P.J. 2010a. Dust: The Inside Story of its Role in the Septemer 11th Aftermath. Lanham, MD: Rowman & Littlefield Publishers.

Lioy, P.J. 2010b. Exposure science: A view of the past and milestones for the future. Environ Health Perspect. 118(8):1081-1090.

Lioy, P.J., and M. Gochfeld. 2002. Lessons learned on environmental, occupational, and residential exposures from the attack on the World Trade Center. Am. J. Ind. Med. 42(6):560-565.

Lioy, P.J., S.S. Isukapalli, L. Trasande, L. Thorpe, M. Dellarco, C. Weisel, P.G. Georgopoulos, C. Yung, M. Brown, S. Alimokhtari, and P.J. Landrigan. 2009. Using national and local extant data to characterize environmental exposures in the National Children's Study: Queens County, New York. Environ Health Perspect. 117(10):1494-1504.

Liu, Y., C.J. Paciorek, and P. Koutrakis. 2009. Estimating regional spatial and temporal variability of PM(2.5) concentrations using satellite data, meteorology, and land use information. Environ. Health Perspect. 117(6):886-892.

Logue, J.M., M.J. Small, D. Stern, J. Maranche, and A.L. Robinson. 2010. Spatial variation in ambient air toxics concentrations and health risks between industrial-influenced, urban, and rural sites. J. Air Waste Manag. Assoc. 60(3):271-286.

Lorber, M., H. Gibb, L. Grant, J. Pinto, J. Pleil, and D. Cleverly. 2007. Assessment of inhalation exposures and potential health risks to the general population that resulted from the collapse of the World Trade Center towers. Risk Anal. 27(5):1203-1221.

Maas, J., R.A. Verheij, P.P. Groenewegen, S. de Vries, and P. Spreeuwenberg. 2006. Green space, urbanity, and health: How strong is the relation? J. Epidemiol. Community Health 60(7):587-592.

Maller, C., M. Townsend, A. Pryor, P. Brown, and L. St Leger. 2006. Healthy nature healthy people: 'Contact with nature' as an upstream health promotion intervention for populations. Health Promot. Int. 21(1):45-54.

Marentette, J.R., K.L. Gooderham, M.E. McMaster, T. Ng, J.L. Parrott, J.Y. Wilson, C.M. Wood, and M. Balshine. 2010. Signatures of contamination in invasive round gobies (*Neogobius melanostomus*): A double strike for ecosystem health? Ecotoxicol. Environ. Saf. 73(7):1755-1764.

Marshall, J.D., R.D. Wilson, K.L. Meyer, S.K. Rajangam, N.C. McDonald, and E.J. Wilson. 2010. Vehicle emissions during children's school commuting: Impacts of education policy. Environ. Sci. Technol. 44(5):1537-1543.

McPherson, E.G., D. Nowak, G. Heisler, S. Grimmond, C. Souch, R. Grant, and R. Rowntree. 1997. Quantifying urban forest structure, function, and value: The Chicago Urban Forest Climate Project. Urban Ecosystems 1(1):49-61.

Meironyte, D., K. Noren, and A. Bergman. 1999. Analysis of polybrominated diphenyl ethers in Swedish human milk. A time-related trend study, 1972-1997. J. Toxicol. Environ. Health 58(6):329-341.

Melia, S., G. Parkhurst, and H. Barton. 2011. The paradox of intensification. Transport. Policy 18(1):46-52.

Miller, K.A., D.S. Siscovick, L. Sheppard, K. Shepherd, J.H. Sullivan, G.L. Anderson, and J.D. Kaufman. 2007. Long-term exposure to air pollution and incidence of cardiovascular events in women. N. Engl. J. Med. 356(5):447-458.

Minckley, W.L., and J.E. Craddock. 1961. Active predation of crayfish on fishes. Prog. Fish Cult. 23(3):120-123.

Mitchell, R., and F. Popham. 2008. Effect of exposure to natural environment on health inequalities: An observational population study. Lancet 372(9650):1655-1660.

Momot, W.T., H. Gowing, and P.D. Jones. 1978. The dynamics of crayfish and their role in ecosystems. Am. Midl. Nat. 99(1):10-35.

NCRP (National Council on Radiation Protection and Measurements). 2008. Uncertainties in the Measurement and Dosimetry of External Radiation. Report No. 158. Bethesda, MD: National Council on Radiation Protection and Measurements.

NCRP (National Council on Radiation Protection and Measurements). 2010a. Radiation Dose Reconstruction: Principles and Practices. Report No. 163. Bethesda, MD: National Council on Radiation Protection and Measurements.

NCRP (National Council on Radiation Protection and Measurements). 2010b. Uncertainties in Internal Radiation Dose Assessment. Report No. 164. Bethesda, MD: National Council on Radiation Protection and Measurements.

Needham, L.L., H. Ozkaynak, R.M. Whyatt, D.B. Barr, R.Y. Wang, L. Naeher, G. Akland, T. Bahadori, A. Bradman, R. Fortmann, L.J. Liu, M. Morandi, M.K. O'Rourke, K. Thomas, J. Quackenboss, P.B. Ryan, and V. Zartarian. 2005. Exposure assessment in the National Children's Study: Introduction. Environ. Health Perspect. 113(8):1076-1082.

NIH (National Institutes of Health). 1985. Report of the National Institutes of Health Ad Hoc Working Group to Develop Radioepidemiologic Tables. NIH Publication No. 85-2748. U.S. Department of Health and Human Services, Public Health Service, National Institutes of Health, Washington, DC [online]. Available: http://www.cdc.gov/niosh/ocas/pdfs/42cfr81/71.pdf [accessed Dec. 13, 2011].

NIOSH (National Institute for Occupational Safety and Health). 2011. Interactive RadioEpidemiological Program. NIOSH-IREP v.5.6 [online]. Available: https://www.niosh-irep.com/irep_niosh/ [accessed Oct. 20, 2011].

NRC (National Research Council). 1983. Risk Assessment in the Federal Government: Managing the Process. Washington, DC: National Academy Press.

NRC (National Research Council). 1984. Assigned Share for Radiation as a Cause of Cancer: Review of Radioepidemiological Tables Assigning Probabilities of Causation: Final Report. Washington, DC: National Academy Press.

NRC (National Research Council). 1994. Science and Judgment in Risk Assessment. Washington, DC: National Academy Press.

NRC (National Research Council). 1999. Strategies to Protect the Health of Deployed U.S. Forces: Force Protection and Decontamination. Washington, DC: National Academy Press.

NRC (National Research Council). 2000a. Strategies to Protect the Health of Deployed U.S. Forces: Analytical Framework for Assessing Risks. Washington, DC: National Academy Press.

NRC (National Research Council). 2000b. Strategies to Protect the Health of Deployed U.S. Forces: Detecting, Characterizing and Documenting Exposures. Washington, DC: National Academy Press.

NRC (National Research Council). 2007. Toxicity Testing in the 21st Century: A Vision and A Strategy. Washington, DC: National Academies Press.

NRC (National Research Council). 2009. Science and Decisions: Advancing Risk Assessment. Washington, DC: National Academies Press.

NRC (National Research Council). 2011. Improving Health in the United States: The Role of Health Impact Assessment. Washington, DC: National Academies Press.

NRC/IOM (National Research Council and Institute of Medicine). 2008. The National Children's Study Research Plan: A Review. Washington, DC: National Academies Press.

O'Fallon, L.R., and A. Dearry. 2002. Community-based participatory research as a tool to advance environmental health sciences. Environ. Health Perspect. 110(suppl. 2):155-159.

Oris, J.T., S. Guttman, A.J. Bailer, J. Reuter, and G. Miller. 2004. Multi-Level Indicators of Ecosystem Integrity in Alpine Lakes of the Sierra Nevada (EPA No. R827643). U.S. EPA Final Project Report, Ecological Indicators Program, 99-NCERQA-E1[online]. Available: http://zoology.muohio.edu/oris/tahoe/SierraFinalReport.htm [accessed May 21, 2012].

Osterauer, R., and H. Koehler. 2008. Temperature-dependent effects of the pesticides thiacloprid and diazinon on the embryonic development of zebrafish (*Danio rerio*). Aquat. Toxicol. 86(4):485-494.

Özkaynak, H., R.M. Whyatt, L.L. Needham, G. Akland, and J. Quackenboss. 2005. Exposure assessment implications for the design and implementation of the National Children's Study. Environ. Health Perspect. 113(8):1108-1115.

Paerl, H.W. 1997. Coastal eutrophication and harmful algal blooms: Importance of atmospheric deposition and groundwater as "new" nitrogen and other nutrient sources. Limnol. Oceanogr. 42(5):1154-1165.

Patel, C.J., and A.J. Butte. 2010. Predicting environmental chemical factors associated with disease-related gene expression data. BMC Med Genomics. 3:17.

Patel, C.J., J. Bhattacharya, and A.J. Butte. 2010. An environment-wide association study (EWAS) on Type 2 Diabetes Mellitus. PLoS ONE 5(5):e10746.

Pickett, S.T.A., M.L. Cadenasso, J.M. Grove, C.H. Nilon, R.V. Pouyat, W.C. Zipperer, and R. Constanza. 2001. Urban ecological systems: Linking terrestrial ecological, physical, and socioeconomic components of metropolitan areas. Annu. Rev. Ecol. Syst. 32:127-157.

PLOTS (Public Laboratory for Open Source Technology). 2012. About PLOTS [online]. Available: http://publiclaboratory.org/about [accessed March 29, 2012].

Rodes, C.E., E.D. Pellizzari, M.J. Dellarco, M.D. Erickson, D.A. Vallero, D.B. Reissman, P.J. Lioy, M. Lippmann, T.A. Burke, and B.D. Goldstein. 2008. ISEA2007 panel: Integration of better exposure characterizations into disaster preparedness for responders and the public. J. Expo. Sci. Environ. Epidemiol. 18(6):541-550.

Rose, K.C., C.E. Williamson, S.G. Schladow, M. Winder, and J.T. Oris. 2009. Patterns of spatial and temporal variability of UV transparency in Lake Tahoe, California-Nevada. J. Geophys. Res. Biogeosci. 114:G00D03.

Schecter, A., M. Pavuk, O. Päpke, J.J. Ryan, L. Birnbaum, and R. Rosen. 2003. Polybrominated diphenyl ethers (PBDEs) in U.S. mother's milk. Environ. Health Perspect. 111(14):1723-1729.

Shankardass, K., R. McConnell, M. Jerrett, J. Milam, J. Richardson, and K. Berhane. 2009. Parental stress increases the effect of traffic-related air pollution on childhood asthma incidence. Proc. Natl. Acad. Sci. USA 106(30):12406-12411.

Sheppard, L., R.L. Prentice, and M.A. Rossing. 1996. Design considerations for estimation of exposure effects on disease risk, using aggregate data studies. Stat. Med. 15(17-18):1849-1858.

Shipler, D.B. 1995. Science in a fishbowl: Public involvement in the Hanford Environmental Dose Reconstruction Project. Fedl. Fac. Environ. J. 6(3):97-108.

Shipler, D.B., B.A. Napier, W.T. Farris, and M.D. Freshley. 1996. Hanford Environmental Dose Reconstruction Project--an overview. Health Phys. 71(4):532-544.

Slikker, W., Jr., M.E. Andersen, M.S. Bogdanffy, J.S. Bus, S.D. Cohen, R.B. Conolly, R.M. David, N.G. Doerrer, D.C. Dorman, D.W. Gaylor, D. Hattis, J.M. Rogers, R. Woodrow Setzer, J.A. Swenberg, and K. Wallace. 2004. Dose-dependent transitions in mechanisms of toxicity. Toxicol. Appl. Pharmacol. 201(3):203-225.

Strauss, W.J., L. Ryan, M. Morara, N. Iroz-Elardo, M. Davis, M. Cupp, M.G. Nishioka, J. Quackenboss, W. Galke, H. Ozkaynak, and P. Scheidt. 2010. Improving cost-effectiveness of epidemiological studies via designed missingness strategies. Stat Med. 29(13):1377-1387.

Su, J.G., M. Jerrett, A. deNazelle, and J. Wolch. 2011. Does exposure to air pollution in urban park shave socioeconomic, racial or ethnic gradients? Environ. Res. 11(3):319-328.

Swift, T.J., J. Perez-Losada, S.G. Schladow, J.E. Reuter, A.D. Jassby, and C.R. Goldman. 2006. A mechanistic clarity model of lake waters: Linking suspended matter characteristics to clarity. Aquat. Sci. 68:1-15.

Szaro, R.C., W.T. Sexton, and C.R. Malone. 1998. The emergence of ecosystem management as a tool for meeting people's needs and sustaining ecosystems. Landscape Urban Plan. 40(1-3):1-7.

Szpiro, A.A., C.J. Paciorek, and L. Sheppard. 2011. Does more accurate exposure prediction necessarily improve health effect estimates? Epidemiology 22(5):680-685.

Teeguarden, J.G., A.M. Calafat, X. Ye, D.R. Doerge, M.I. Churchwell, R. Gunawan, and M.K. Graham. 2011. Twenty-four hour human urine and serum profiles of bisphenol A during high-dietary exposure. Toxicol. Sci. 123(1):48-57.

Thomas, D.C. 1988. Models for exposure-time-response relationships with applications to cancer epidemiology. Annu. Rev. Publ. Health 9:451-482.

Thomas, D. 2010. Gene-environment-wide association studies: Emerging approaches. Nat. Rev. Genet. 11(4):259-272.

Tucker, A.J., C.E. Williamson, K.C. Rose, J.T. Oris, S. Connelly, M.H. Olson, and D.L. Mitchell. 2010. Ultraviolet radiation affects invasibility of lake ecosystems by warmwater fish. Ecology 91(3): 882-890.

Tucker, A.J., C.E. Williamson, and J.T. Oris. in press. Development and application of a UV attainment threshold for the prevention of warmwater aquatic invasive species. Biol. Invasions in press.

Tulve, N.S., L.S. Sheldon, and R.C. Fortmann. 2010. Workshop on Optimizing Exposure Metrics for the National Children's Study: Summary of Workgroup Discussions and Recommendations. EPA/600/R-10/064. Office of Research and Development, U.S. Environmental Protection Agency, Research Triangle Park, NC. June 2010 [online]. Available: http://oaspub.epa.gov/eims/eimscomm.get file?p_download_id=497487 [accessed Jan. 2, 2012].

Wang, R. Y., L.L. Needham, and D.B. Barr. 2005. Effects of environmental agents on the attainment of puberty: Considerations when assessing exposure to environmental chemicals in the National Children's Study. Environ. Health Perspect. 113(8):1100-1107.

Wiener, J.G., D.P. Krabbenhoft, G.H. Heinz, and A.M. Scheuhammer. 2002. Ecotoxicology of mercury. Pp. 409-463 in Handbook of Ecotoxicology, 2nd Ed., D.J. Hoffman, B.A. Rattner, G.A. Burton, and J. Cairns, eds. Boca Raton: Lewis.

Wier, M., J. Weintraub, E. Humphreys, E. Seto, and R. Bhatia. 2009a. An area-level model of vehicle-pedestrian injury collisions with implications for land use and transportation planning. Accid. Anal. Prev. 41(1):137-145.

Wier, M., C. Sciammas, E. Seto, R. Bhatia, and T. Rivard. 2009b. Health, traffic, and environmental justice: Collaborative research and community action in San Francisco, California. Am. J. Public Health 99(suppl. 3):S499-S504.

Wolch, J., M. Jerrett, K. Reynolds, R. McConnell, R. Chang, N. Dahmann, K. Brady, F. Gilliland, J.G. Su, and K. Berhane. 2011. Childhood obesity and proximity to urban parks and recreational resources: A longitudinal cohort study. Health Place 17(1):207-214.

Zablotska, L.B., J.P. Ashmore, and G.R. Howe. 2004. Analysis of mortality among Canadian nuclear power industry workers after chronic low-dose exposure to ionizing radiation. Radiat. Res. 161(6):633-641.

4

Demands for Exposure Science

INTRODUCTION

Knowledge of exposure is key to predicting, preventing, and reducing environmental and human health risks. A robust exposure science is necessary to support policy decisions for managing potentially harmful exposures without adversely affecting economic activities, personal liberties, and the health of people. The need for exposure science extends beyond policy considerations, however, and includes societal goals related to population health, economic security, and human well-being. This chapter addresses the demands for exposure science that the committee's proposed vision will help to meet. The committee's vision will help transform exposure science into a more forward-looking discipline that supports universal exposure surveillance and integrated predictive systems that facilitate early detection of and even anticipate harmful exposures.

The committee's vision (Chapter 2) arises in part from multiple and complex scientific, societal, commercial, and policy demands. The committee did not attempt to develop an exhaustive list of those demands but rather selected examples to illustrate their nature and their importance in shaping research needs for exposure science in the 21st century. This chapter builds on the concepts and terminology in Chapter 1 and the applications of exposure science in Chapter 3. It sets the stage for Chapter 5, which identifies scientific and technologic advancements needed to support the committee's vision.

The committee broadly explored research-based and decision-based activities to identify emerging needs for exposure information. This exploration reveals that the demand for exposure information is growing. One example of this is the knowledge gap resulting from the introduction of thousands of new chemicals into the market each year. The U.S. Toxic Substances Control Act and the Green Chemistry Initiative of the California Department of Toxic Substances Control (CA DTSC 2008) demonstrate that the rate of introduction of new substances exceeds our ability to design and conduct exposure assessments of these new chemicals and their mixtures that enter the market. Other examples of emerging demands for exposure information are EPA's Premanufacturing Notice requirements and the European Union's program for the Registration, Evaluation, Authorization and Restriction of Chemicals. The industries that market chemicals and the government agencies that regulate them need more and better exposure information to conduct screening and regulatory assessments.

Another example is the increasing need to address long-term health effects of low-level exposures to chemical, biologic, and physical stressors over years or decades, such as low-level radiation exposure. A dearth of exposure data contributed to uncertainty in communicating the radiation risk posed by the Fukushima incident in Japan to policy-makers and the public. Previous opportunities to reduce uncertainties through the collection of more and better exposure data have been missed, including opportunities in the aftermath of the Soviet Union's April 1986 Chernobyl nuclear incident, which spewed radionuclides over a large swath of Europe (Normile 2011). There were few systematic or sustained applications of exposure-science techniques in the collection of radiation-exposure data at Chernobyl (UNSCEAR 2011).

Growing efforts to collect, organize, and evaluate medical-surveillance data in the absence of corresponding efforts to assemble, evaluate, and track exposure data present another example of the need for data. The paucity of exposure data has been observed repeatedly—in the followup of health effects in veterans of the Gulf War (IOM 2000), in the Centers for Disease Control and Prevention (CDC) Health Tracking Program (CDC 2011a), and in the monitoring of the health of volunteers and professionals after exposures to the 2010 Gulf Oil Spill (IOM 2010; King and Gibbons 2011).

The committee defined the complex and overlapping needs for exposure information in four broad categories: health and environmental science, market, societal, and policy and regulatory (illustrated schematically in Figure 4-1). Health and environmental sciences require reliable quantitative data on human and ecosystem exposures. Market demands require the identification and control of exposures resulting from the manufacture, distribution, and sale of products and the provision of services (for example, energy, transportation, and health care). Societal demands arise from the aspirations of individuals and communities—relying on an array of health, safety, and sustainability information—for example, to maintain local environments, personal health, the health of workers who make consumables, and the health of the global environment. Policy-makers drive the need for exposure science when they require knowledge to inform their actions—particularly the setting of policies directed at mitigating environmental risks and avoiding hazards in cost-effective ways. Policy-makers need to establish a balance among the different science (health), market, and societal demands as they establish regulations and set budgetary priorities. The remainder of this chapter explores the four categories of needs for exposure science information. The committee recognizes that many of these demands can conflict. For example, individuals and communities may have different goals

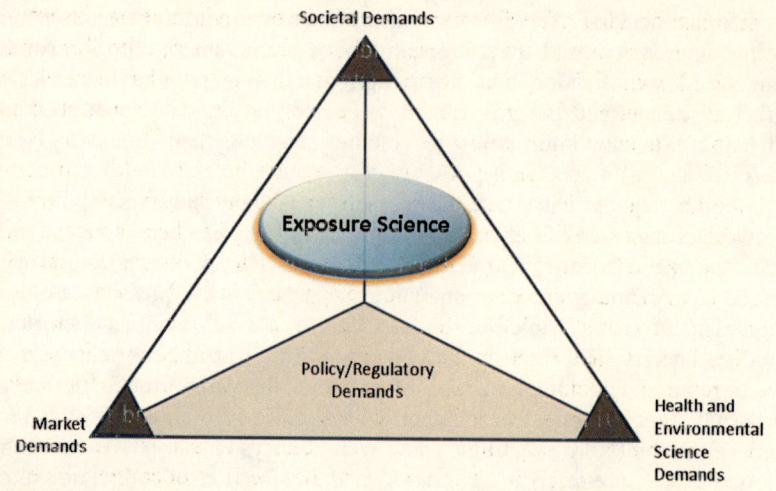

FIGURE 4-1 The four major demands for exposure science.

and aspirations with respect to research and policies to maintain the environment, the health of individuals, and communities. Similarly, policy makers and regulatory agencies often have different and even competing goals, and each will be interested in exposure studies that support their particular perspectives. The goal is to explore the various demands, recognizing the potential for competing and conflicting demands.

HEALTH AND ENVIRONMENTAL SCIENCE DEMANDS

The need to protect human health has been and will continue to be an important demand for exposure information (NRC 2009a). Accurate assessment of human exposures is a critical component of environmental health research (McKone et al. 2009). Air pollution epidemiology, risk assessment, health tracking, and accountability assessments are examples of health research studies that require but often lack adequate exposure information (McKone et al. 2009). The expanding number of environmental factors that are or will be the focus of health research creates a continuing demand for exposure information. In many health studies the lack of accurate exposure information has led to the use of questionnaires and qualitative assessments in place of more robust quantitative observations.

Demand for health and environmental science information includes the need for more and improved data on broad issues, such as direct stressor–target relationships—for example, air pollution and health and the multiple, complex, and sometimes indirect linkages among environmental exposures and ecosystems, water and land resources, and the built environment. Demands for exposure information also arise from specific health or environmental issues—for example, a rise in autism, asthma, or childhood brain cancer; reproductive failure in specific wildlife populations; or deterioration of popular local habitats. There is also a need to integrate health and ecologic sciences to support a more harmonized framework for assessing the fate and effects of industrially produced and naturally occurring pollutants. Protecting human health and ecosystem integrity requires long-term and spatially and temporally resolved tracking of multiple stressors. For humans, that type of tracking has been conceptualized as the "exposome", defined as collective exposures from conception on (Wild 2005, 2012). An analogous ecologic-science approach is the National Ecological Observatory Network (NEON)—a continental-scale research instrument proposed by the National Science Foundation (NSF) to provide a nationally networked research, communication, and informatics infrastructure for biologic systems (NEON 2011)—intended to assess the direct effects and feedbacks between environmental change and biologic processes. A similar NSF program, the Long Term Ecological Research network, has been in operation since 1980 (LTER 2012).

Environmental exposures contribute substantially to human health risks, accounting for a greater fraction of risk than genetic variation (Rappaport and Smith 2010). New analytic capabilities are needed for environmental surveillance and biomonitoring and for linking biomarkers to stressors on the basis of pharmacokinetic and pharmacodynamic models (Sohn et al. 2004; Clewell et al. 2008).

Health

With regard to the influence of environmental exposures on health, direct cause–effect relationships are sometimes apparent, as in the cases of radiation from nuclear weapons in Japan; dioxin in Seveso, Italy; and methyl isocyanate in Bhopal, India. Such relationships can be evident even if there is a substantial lag in the development of health effects, as in the 10- to 50-year delay in development of mesothelioma from asbestos exposure.

More commonly, however, exposure–effect relationships are difficult to establish because of other variables. For example, studies of cancer risk in migrants show that environmental factors can cause large increases in risk (see Ziegler et al. 1993). The increase in risk is attributed to lifestyle differences, such as differences in air and water, food, pharmaceuticals, and many household and occupational exposures. Effective monitoring methods, such as CDC's National Health and Nutrition Examination Survey (NHANES), directly reveal such health changes as the rise in obesity. An ability to link those changes rapidly to specific exposures—for example, endocrine-disrupting chemicals (Heindel 2003), diet, and urban patterns—requires continuing exposure assessment. The need for innovative and cost-effective means of separating and measuring specific exposures constitutes an important demand for exposure science.

Scientific advances in epigenomic research (the study of epigenetic modification at a level much larger than a single gene) have revealed long-term effects of early-life exposures on modification of DNA-methylation patterns (Jirtle and Skinner 2007). Relevant exposures and vulnerable life stages are beginning to be understood, but preliminary results illustrate the need for better exposure data for assessing long-term disease risks. That need is underscored by concern about transgenerational risks posed by fetal exposures, including those during ovarian-cell development, which can affect health outcomes in later generations (Skinner et al. 2010). These types of observations suggest the opportunity for novel approaches to translate internal markers into measures of exposure at critical life stages. To achieve this, scientists need to quantify current exposures and preserve the data in forms that permit them to be used in the future to elucidate transgenerational risks posed by particular exposures.

Increasing use of burden of disease metrics (such as disability adjusted life years [DALYs]) and comparative risk assessment covering a wide array of risk factors, including chemical exposures, diet, and lifestyle, to inform policy decisions demands better and more consistent methods of characterizing diverse exposures in large populations. The problem that arises from this demand is the need to provide measures of environmental exposures that are consistent in statistical and causal terms with measures used to characterize exposures to nonenvironmental risk factors—such as smoking, unsafe sex, and micronutrient deficiencies. Consistent measures of both environmental and nonenvironmental exposures are needed if meaningful policy comparisons are to be made.

Environment

There are growing demands for comprehensive information on global, regional, and local environmental problems. Improvements in air and water quality, mostly in developed countries, have been made possible by advances in science and technology. Those improvements will provide a foundation for addressing future demands stemming from growing populations and shrinking resources.

Air Quality

Over the last 2 decades, emissions from energy use in transportation, power generation, industry, and households have steadily decreased in developed countries as a result of emission-control strategies (HEI 2010; EPA 2011a,b; NRC 2010). This has contributed to decreased ambient concentrations of particulate and gaseous air pollutants in many cities, and the effects of transported emissions from other states and countries have become important contributors to total exposures. Those changes drive new needs to monitor fluctuations in ambient air pollutants in space and time, link them to mitigation strategies, and assess health benefits of reducing human exposures. That will require development and validation of spatiotemporal exposure models that will use data on land use, human activity, housing characteristics, atmospheric concentrations, and personal monitoring. The models need to be specific to different populations, especially populations that are particularly susceptible, such as children and people who have pulmonary and cardiovascular diseases.

Water Quality

Water-quality and water-quality impact assessments have changed substantially over the last several decades with a greater focus on understanding the complex interactions among human populations and water supplies. This focus has created a growing demand for water-pathway exposure assessments. The number and quantities of new chemicals and materials (such as nanomaterials) now found in waste streams far exceed our capacity to monitor them (Kim et al. 2010; Nowack 2010). An ongoing need is to evaluate and limit adverse water-quality effects on aquifers, waterways, forests, and agricultural lands. Improved data on regional and global distribution of persistent chemicals of the types monitored in air-quality studies are needed to address these critical issues. In addition, although cumulative effects of mixtures are largely unknown, there is concern that global accumulations of contaminant mixtures may result in unexpected long-term effects on human and ecologic targets and on the water resources themselves (Macdonald et al. 2000; Woodruff et al. 2011).

Global Climate Change

Global climate change is expected to bring increasingly frequent extreme weather and local environmental changes that have the potential to affect human health, ecologic health, and key resources in several direct and indirect ways (Patz et al. 2007; NRC 2009a; USGCRP 2009). The effects will include those from increased temperature, such as acute and chronic health effects; those from extreme weather, such as physical injuries and drownings, structural collapses, and declines in habitability due to mold and other kinds of contamination; and indirect effects, such as shortages of clean water and increasing concentrations of contaminants due to drought (Frumkin et al. 2008). The National Research Council report on global climate change and human health (NRC 2009a) and a U.S. Council on Environmental Quality Climate Change Adaptation Task Force report (CEQ 2010) addressed recommendations for protecting against those effects. Global climate change will bring new needs for exposure science to examine the effects of climate changes on exposures to new and altered chemical, physical, and biologic stressors. Programs to address climate change and health have been established in CDC (CDC 2011b) and the National Institute for Environmental Health Sciences (NIEHS 2011). Those programs are seeking more input from the exposure-science community; see, for example, the CDC national conversation with its emphasis on public health and chemical exposures (Brown 2011; CDC 2011c).

Energy Demand

The production and use of energy emit pollutants that have been linked to diseases (IIASA 2011) through exposure in the ambient environment (for example, to power-plant emissions) (NRC 2010) and in the indoor environment (for example, to cooking and heating emissions) (Smith et al. 2004). NRC (2010) reported that the quantifiable public-health costs of all energy production, distribution, and use in 2005 totaled $120 billion and were due mostly to criteria air pollutants. Of that amount, $62 billion was attributable to electricity (mainly coal) and about $56 billion to transportation; the remainder was attributable to process heat (for example, industrial boilers) and comfort control (for example, home or commercial-building heating and air-conditioning systems). There are expected to be increases in energy use, but an additional demand for exposure science will occur as a result of transitions from one energy source to another. Energy sources have different effects throughout the use chain (or fuel life cycle), from resource capture through energy production to conversion, distribution, and end use. Because the full burden of extant energy systems has not been adequately characterized (IIASA 2011), there is no appropriate baseline against which to compare the relative benefits of new systems. As world leaders consider options for changing the portfolio of future energy sources, there is growing demand for assessments of effects associated with the various options, including pollutant exposures, and a need to develop strategies to minimize the effects.

Sustainability

Sustainability describes both a process to ensure and a goal of ensuring long-term human well-being and ecologic health (NRC 2011). All technologies have benefits and effects, and an important aspect of long-term sustainability is that technologies achieve an overall balance. Increasing use of technology assessment can be expected to avoid strategic errors that could derail a promising technology and improve policy decisions to avoid long-term adverse effects. Life-cycle assessment of all stages of a process—including-raw material extraction, manufacturing, distribution, use, and disposal—is an accepted approach to evaluating resource consumption and resulting pollution (Guinée et al. 2010). Such analysis is critical for supporting decision-making (Guinée and Heijungs 1993) to ensure sustainability of the environment and resources and to assess health and ecologic effects. NRC (2011) recommends that the Environmental Protection Agency (EPA) develop a "sustainability toolbox" that collectively makes it possible to analyze present and future consequences of alternative decision options on the full array of social, environmental, and economic indicators.

Because of increasing demands for sustainability metrics, an associated demand for exposure-science surveillance and for predictive tools to support life-cycle analysis can be anticipated (see, for example, the USEtox model [USEtox.org 2009]).

MARKET DEMANDS

Industries and investors want to limit their liability for health and environmental damages and minimize regulatory oversight, and the rapid increase in technologic applications in commerce has created market demands for exposure science that often correspond with demands for information on health and the environment. For example, industries and investors use electronic media as a means of promoting and assessing consumption of their products, considering profitability, regulations, and liability. Organizations and Web sites provide health and environmental information and product scores or rankings of a wide variety of consumer products (for example, GoodGuide 2011). Social networks provide tools for building and exchanging information about the health, environmental, and societal effects of consumer products. Organizations, activities, and tools encourage consumers to consider alternative products and behaviors that can reduce such effects. Consequently, market demands for exposure science include the need for better and more extensive insight into how human activities, including consumption habits, contribute to pollutant emissions and how the emissions contribute to human and ecologic exposures.

Growth in Consumption and Demand for Sophisticated Consumer Products

The increase in global, national, and individual purchases of consumer products places a substantial burden on environmental resources (NRC 2011) and could continue to for decades before stabilization at a sustainable level (Hertwich 2005). Many factors affect that burden, such as increased population, increased personal wealth, and reductions in the useful lifetime of consumer products. Technologic advances have fueled an expectation for improved products, and that expectation contributes to their relatively short lifetimes (for example, mobile telephones, video players, and computers). New products with enhanced properties also lead to replacement of well-functioning equivalent products (for example, more efficient light bulbs or programmable appliances replace existing light bulbs or appliances). Studies of trends in consumer products show that the stability and durability of products have important roles in exposure potential (Hertwich 2005). Stable and durable products resist degradation and contribute less to emissions during their lifetime, but their disposal can be problematic. Information on exposure potential of new products is needed for evaluation of long-term health footprints of exposures and associated risks. There is also a need for exposure data to guide policy in the development of short-lived products and new products designed to generate premature product replacement. Consumers can play an increasing role by demanding information on the effects of products on the climate, resources, and health—including considerations about exposures across the life-cycle of products.

Food Supply

Exposure science provides critical input for ensuring the sustainability and safety of the food supply. With increasing population growth and internationalization of the food supply, there is expected to be an increasing demand for exposure science (Schmidhuber and Shetty 2005). Demographics of health and disease suggest that diet is a major source of environmental exposures (Ames 1983; Willett 2002), and the increasing globalization and consolidation of food-distribution networks have created a potential for rapid, widespread dissemination of contaminated products (Regattieri al. 2007). In light of economic, political, and nutritional advantages of local food production, that has led to programs to reverse trends in the globalization of the food supply. The competing trend toward globalization vs localization will demand an expansion of exposure-surveillance structures to manage and monitor the changing array of agrarian practices and their influence on environmental quality (NRC 2002). The far-reaching effects of the changes will require a critical role for exposure science to support evaluation and development of policies.

Green Chemistry

Green chemistry or, more broadly, green commerce includes the design of products and processes with a focus on sustainability with regard to resource consumption and energy use, often accompanied by an effort to limit the human and ecologic health footprint (CA DTSC 2008).[1] Green commerce encounters the same challenges as other businesses as practical considerations of profitability often require use of more available resources according to supply and demand; that is, as the feedstock diminishes and prices rise, manufacturers seek alternatives. Small businesses are especially vulnerable to such variations and could benefit from publicly accessible exposure data. New approaches in exposure assessment and information dissemination are critical for decreasing the pollution footprint of products and services while allowing for adaptive responses by free enterprise.

---

[1] EPA describes green-chemistry goals as including source reduction and the prevention of chemical hazards, such as through the use of feedstocks and reagents that are less hazardous to human health and the environment, the design of syntheses and other processes to be less energy-intensive and material-intensive, the use of feedstocks derived from annually renewable resources or from abundant waste, the design of chemical products for increased and easier reuse or recycling, treatment to render chemicals less hazardous, and proper disposal of chemicals (EPA 2011a).

SOCIETAL DEMANDS

The health, environmental, and market demands discussed above are direct reflections of a society and the complex needs and desires of its constituents. Over the last century, there has been a dramatic rise in world population combined with an increase in urbanization, and the result has been profound changes in not only where people live but how they live. The evolution of the U.S. and world populations from primarily agrarian communities to megacities and sprawling suburbs has led to societal (and scientific) questions about effects on human health and well-being and on ecosystems. Concurrently, there have been dramatic changes in what is eaten and how, how and how often people travel, and how technologies are used for communication. Major societal demands for exposure science include the understanding and assessment of the effects of urbanization and urban land-use modifications and of changes in manufacturing, consumption, and transportation.

Urbanization and Land-Use Changes

Projections of changes in urbanization and land use indicate both increased need for and more systematic means of exposure surveillance in the coming decades. The density of economic activity increases with urban population density (World Bank 2009); half the world population now lives in cities, and the UN projects that about 75% will live in cities by 2050 (UN 2008). The growth of suburban areas is especially pronounced in North America, where cities are highly energy-intensive and transportation is dominated by automobiles, but similar patterns of suburbanization are evident in many other places. Increased automobile use contributes to environmental problems—increased air pollution, storm-water runoff due to impervious surfaces, and higher intake of pollutants when roadways are near residences (Hough 1995; Frumkin et al. 2004; Jerrett et al. 2010).

There is a need for a systematic process to evaluate exposures associated with different layouts and designs of urban areas to guide planning to optimize urban and suburban structures with consideration of exposures and health (Marshall et al. 2005). The factors that have increased exposures appear to be growing as many previously poor nations undergo rapid economic transformations and higher economic growth (Wang et al. 2005). The complexity of those driving forces and the need to protect human and ecosystem health can be met with increased and more systematic exposure surveillance in the coming decades of the 21st century.

Societal Issues in Manufacturing and Transportation

Economic changes in the developing and developed worlds have altered where and how products are produced and distributed. There is an expectation that consumer goods are to be convenient, mobile, and accessible. Rapid changes in style and capacity (for example, of telephones) lead to quick product turnover, which results in a growing waste-management challenge. The production of steel, electronics, and many consumer products is moving to Asia, leaving the large industrial regions of Europe and North America seeking "rebirth" and risking decay and abandonment. Transportation systems are facing growing demands. Web-based purchasing creates a growing need for home-delivery networks. Both tourists and business travelers are taking more and longer airline flights. Those dramatic changes in the world of manufacturing and distribution give rise to concerns and questions about how they will affect humans and the environment, motivating the acquisition of information on changing exposure patterns.

Other Societal Concerns

There are many other societal concerns that demand accurate and more comprehensive exposure information. For example exposures to biologic stressors in water supplies and food. In urban areas there are significant concerns about exposure to noise, which is often only well monitored and researched near "hot spots" such as airports and major roadways. Mixed exposures among chemical, physical, and biologic stressors are also of concern, but difficult to track and evaluate (WHO 2012).

For example, studies in the European Union reveal that excessive noise can harm human health and interfere with people's daily activities. It can disturb sleep, cause cardiovascular and psychologic stress, reduce performance, and cause changes in behavior (WHO 2012). Addressing these health concerns requires more reliable monitoring of noise levels over a broad range of geographic areas.

POLICY AND REGULATORY DEMANDS

Exposure science used in policy-making can provide information to support environmental protection, resource management, chemical regulation, manufacturing goals, and health, energy, climate, and economic policies. The policy and regulatory demands for exposure science are unique in their link to governments. Policy-makers need to make tradeoffs among a broad array of outcome options. For example, they use exposure information to address conflicting societal, commercial, and scientific considerations, and they use it to monitor the health and environmental benefits of regulations (NRC 2007). Policy-makers have the capacity to use exposure information to inform and motivate activities or to address the reluctance of other policy-making entities or others to take action. For example, robust exposure science is a key asset in an era of limited resources. It is particularly useful for an agency that has responsibility for promoting the health and sustainability of communities in separating perceived effects and benefits based on anecdotal evidence from those which are large and well documented, steering limited resources away from ineffective interventions. (The exposure metric called "Intake Fraction" is an example of a tool that might be used in regulations to improve health protection, see Chapter 1 discussion.)

Policy-makers and regulators have a demand for exposure information to inform the concerned public about products and exposures, to establish emergency preparedness and response, to set priorities for research and regulation among chemicals or stressors of concern, and to allocate funding and set policies for managing knowledge-integration systems to address health and ecosystem protection. Adding to the policy demands for exposure science are the community demands for access to technologies that allows community members to work with scientists, to generate their own exposure data, and to more effectively participate in the environmental policy and regulatory processes (Brown et al. 2012).

BUILDING CAPACITY TO MEET DEMANDS

Health and environmental, market, societal, and policy and regulatory demands are creating increased needs for exposure science in the 21st century. Meeting those needs will require a scientific framework that supports the development of technologies to collect, analyze, and integrate exposure-science data. The remainder of this report addresses the framework for building the capacity to meet the demands for exposure science in the 21st century.

REFERENCES

Ames, B.N. 1983. Dietary carcinogens and anticarcinogens. Oxygen radicals and degenerative diseases. Science 221(4617):1256-1264.
Brown, V.J. 2011. Are we on the same page? Action agenda of the National Conversation on Public Health and Chemical Exposures. Environ. Health Perspect. 119(11): A484-A487.
Brown, P., J. Green Brody, R. Morello-Frosch, J. Tovar, A.R. Zota, and R.A. Rudel. 2012. Measuring the success of community science: The northern California household exposure study. Environ. Health Perspect. 120(3):326-331.
CA DTSC (California Department of Toxic Substances Control). 2008. Green Chemistry Initiative Final Report. California Department of Toxic Substances Control. December 2008 [online]. Available: http://www.dtsc.ca.gov/PollutionPrevention /GreenChemistryInitiative/upload/GREEN_Chem.pdf [accessed Dec. 27, 2011].
CDC (Centers for Disease Control and Prevention). 2011a. National Environmental Public Health Tracking Program [online]. Available: http://www.cdc.gov/nceh/track ing/ [accessed Dec. 10, 2011].
CDC (Centers for Disease Control and Prevention). 2011b. Climate and Health Program [online]. Available: http://www.cdc.gov/climatechange/ [accessed Sept. 15, 2011].

CDC (Centers for Disease Control and Prevention). 2011c. National Conversation on Public Health and Chemical Exposures [online]. Available: http://www.nationalconversation.us/ [accessed Dec. 27, 2011].

CEQ (U.S. Council on Environmental Quality). 2010. Progress Report of the Interagency Climate Change Adaptation Task Force: Recommended Actions in Support of a National Climate Change Adaptation Strategy, October 5, 2010 [online]. Available: http://www.whitehouse.gov/sites/default/files/microsites/ceq/Interagency-Climate-Change-Adaptation-Progress-Report.pdf [accessed Dec. 27, 2011].

Clewell, H.J., Y.M. Tan, J.L. Campbell, and M.E. Andersen. 2008. Quantitative interpretation of human biomonitoring data. Toxicol. Appl. Pharmacol. 231(1):122-133.

EPA (U.S. Environmental Protection Agency). 2011a. Introduction to the Concept of Green Chemistry. Office of Chemical Safety and Pollution Prevention, Office of Pollution Prevention and Toxics, U.S. Environmental Protection Agency [online]. Available: http://www.epa.gov/gcc/pubs/about_gc.html [accessed Dec. 27, 2011].

EPA (U.S. Environmental Protection Agency). 2011b. National Emission Inventory Data & Documentation: 1999 through 2008. Technology Transfer Network Clearinghouse for Inventories and Emissions Factors, U.S. Environmental Protection Agency [online]. Available: http://www.epa.gov/ttnchie1/net/ [accessed Dec. 27, 2011].

Frumkin, H., L. Frank, and R.J. Jackson. 2004. Urban Sprawl and Public Health: Designing, Planning, and Building for Healthy Communities. Washington, DC: Island Press.

Frumkin, H., J. Hess, G. Luber, J. Malilay, and M. McGeehin. 2008. Climate change: The public health response. Am. J. Public Health. 98(3):435-445.

Good Guide. 2011. Find Healthy, Green, Ethical Products According to Scientific Ratings. Good Guide, Inc. [online]. Available: http://www.goodguide.com/ [accessed May 16, 2011].

Guinée, J., and R. Heijungs. 1993. A proposal for the classification of toxic substances within the framework of life cycle assessment of products. Chemosphere 26(10):1925-1944.

Guinée, J.B., R. Heijungs, G. Huppes, A. Zamagni, P. Masoni, R. Buonamici, T. Ekvall, and T. Rydberg. 2010. Life cycle assessment: Past, present, future. Environ. Sci. Technol. 45(1): 90-96.

HEI (Health Effects Institute). 2010. Traffic-Related Air Pollution: A Critical Review of the Literature on Emissions, Exposure, and Health Effects. Special Report 17. Boston, MA: Health Effects Institute [online]. Available: http://pubs.healtheffects.org/view.php?id=334 [accessed Dec. 12, 2011].

Heindel, J.J. 2003. Endocrine disruptors and the obesity epidemic. Toxicol. Sci. 76(2):247-249.

Hertwich, E.G. 2005. Life cycle approaches to sustainable consumption: A critical review. Environ. Sci. Technol. 39(13):4673-4684.

Hough, M. 1995. Cities and Natural Process, 2nd Ed. New York: Routledge.

IIASA (International Institute for Applied Systems Analysis). 2011. Global Energy Assessment, Chapter 4. Energy and Health. Cambridge University Press.

IOM (Institute of Medicine). 2000. Gulf War and Health, Volume 1. Depleted Uranium, Sarin, Pyridostigmine Bromide, Vaccines. Washington, DC: National Academy Press.

IOM (Institute of Medicine). 2010. Research Priorities for Assessing Health Effects from the Gulf of Mexico Oil Spill: A Letter Report. Washington, DC: National Academies Press.

Jerrett, M., S. Gale, and C. Kontgis. 2010. An environmental health geography of risk. Pp. 418-445 in A Companion to Health and Medical Geography, T. Brown, S. McLafferty, and G. Moon, eds. Chichester: Wiley-Blackwell.

Jirtle, R.L., and M.K. Skinner. 2007. Environmental epigenomics and disease susceptibility. Nat. Rev. Genet. 8(4):253-262.

Kim, B., C.S. Park, M. Murayama, and M.F. Hochella. 2010. Discovery and characterization of silver sulfide nanoparticles in final sewage sludge products. Environ. Sci. Technol. 44(19):7509-7514.

King, B.S., and J.D. Gibbins. 2011. Health Hazard Evaluation of Deepwater Horizon Response Workers. Health Hazard Evaluation Report HETA 2010-0115&2010-0129-3138. Centers for Disease Control and Prevention, National Institute for Occupational Safety and Health. August 2011. Available: http://www.cdc.gov/niosh/hhe/reports/pdfs/2010-0115-0129-3138.pdf [accessed Dec. 27, 2011].

LTER (The Long Term Ecological Research). 2012. 2011 Strategic and Implementation Plan (SIP). The Long Term Ecological Research Network [online]. Available: http://www.lternet.edu [accessed April 30, 2012].

Macdonald, R.W., L.A. Barrie, T.F. Bidleman, M.L. Diamond, D.J. Gregor, R.G. Semkin, W.M. Strachan, Y.F. Li, F. Wania, M. Alaeee, L.B. Alexeeva, S.M. Backus, R. Bailey, J.M. Bewers, C. Gobeil, C.J. Halsall, T. Harner, J.T. Hoff, L.M.M. Jantunen, W.L. Lockhart, D. Mackay, D.C. Muir, J. Pudykiewicz, K.J. Reimer, J.N. Smith, and G.A. Stern, 2000. Contaminants in the Canadian Arctic: 5 years of progress in understanding sources, occurrence and pathways. Sci. Total Environ. 254 (2-3): 93-234.

Marshall, J.D., S.K. Teoh, and W.W. Nazaroff. 2005. Intake fraction of nonreactive vehicle emissions in U.S. urban areas. Atmos. Environ. 39(7):1363-1371.

McKone, T.E., P.B. Ryan, and H. Ozkaynak. 2009. Exposure information for ozone, PM and toxic air pollutants in environmental health research: Current opportunities and future directions. J. Expo. Sci. Environ. Epidemiol. 19(1):30-44.

NEON, Inc. 2011 Science Strategy: Enabling Continental-Scale Ecological Forecasting. National Ecological Observatory Network [online]. Available: http://www.neoninc.org/sites/default/files/NEON_Strategy_2011u2.pdf [accessed Dec. 28, 2011].

NIEHS (National Institute of Environmental Health Sciences). 2011. Climate Change and Human Health: Programs and Initiatives. National Institute of Environmental Health Sciences, National Institute of Health [online]. Available: http://www.niehs.nih.gov/about/od/programs/climatechange/index.cfm [accessed Sept. 15, 2011).

Normile, D. 2011. Fukoshima revives the low-dose debate. Science 332(6032):908-910.

Nowack, B. 2010. Nanosilver revisited downstream. Science 330(6007):1054-1055.

NRC (National Research Council). 2002. Frontiers in Agricultural Research: Food, Health, Environment, and Communities. Washington, DC: National Academies Press.

NRC (National Research Council). 2007. Scientific Review of the Proposed Risk Assessment Bulletin from the Office of Management and Budget. Washington, DC: National Academies Press.

NRC (National Research Council). 2009a. Restructuring Federal Climate Research to Meet the Challenges of Climate Change. Washington, DC: National Academies Press.

NRC (National Research Council). 2010. Hidden Costs of Energy: Unpriced Consequences of Energy Production and Use. Washington, DC: National Academies Press.

NRC (National Research Council). 2011. Sustainability and the U.S. EPA. Washington, DC: National Academies Press.

Patz, J.A., H.K. Gibbs, J.A. Foley, J.V. Rogers, and K.R. Smith. 2007. Climate change and global health: Quantifying a growing ethical crisis. EcoHealth 4(4):397-405.

Rappaport, S.M., and M.T. Smith. 2010. Environment and disease risks. Science 330(6003):460-461.

Regattieri, A., M. Gamberi, and R. Manzini. 2007. Traceability of food products: General framework and experimental evidence. J. Food Eng. 81(2):347-356.

Schmidhuber, J., and P. Shetty. 2005. The Nutrition Transition to 2030: Why Developing Countries are Likely to Bear the Major Burden. Presentation at the 97th Seminar of the European Association of Agricultural Economists, April 21-22, 2005, University of Reading, UK [online]. Available: http://www.fao.org/fileadmin/te mplates/esa/Global_persepctives/Long_term_papers/JSPStransition.pdf [accessed May 30, 2012].

Skinner, M.K., M. Manikkam, and C. Guerrero-Bosagna. 2010. Epigenetic transgenerational actions of environmental factors in disease etiology. Trends Endocrinol. Metab. 21(4):214-222.

Smith, K.R., S. Mehta, and M. Maeusezahl-Feuz. 2004. Indoor smoke from household solid fuels. Pp. 1435-1493 in Comparative Quantification of Health Risks: Global and Regional Burden of Disease due to Selected Major Risk Factors, Vol. 2, M. Ezzati, A.D. Lopez, A.D. Rodgers, and C.J.L. Murray, eds. Geneva: World Health Organization.

Sohn, M.D, T.E. McKone, and J.N. Blancato. 2004. Reconstructing population exposures from dose biomarkers: Inhalation of trichloroethylene (TCE) as a case study. J. Expo. Anal. Environ. Epidemiol. 14(3):204-213.

UN (United Nations). 2008. An Overview of Urbanization, Integral Migration, Population Distribution and Development in the World. UN/POP/EGM-URB/2008/01, January 14, 2008. Contributed Paper at United Nations Expert Group Meeting on Population Distribution, Urbanization, Internal Migration and Development, January 21-23, New York. U.N. Department of Economic and Social Affairs, Population Division [online]. Available: http://www.un.org/esa/population/meet ings/EGM_PopDist/P01_UNPopDiv.pdf [accessed Dec. 10, 2011].

UNSCEAR (United Nations Scientific Committee on the Effects of Atomic Radiation). 2011. Sources and Effects of Ionizing Radiation. UNSCEAR 2008 Report to the General Assembly, Vol. II, Annex D. Health Effects Due to Radiation from the Chernobyl Accident. New York: United Nations Press [online]. Available: http://www.unscear.org/docs/reports/2008/11-80076_Report_2008_Annex_D.pdf [accessed Dec. 27, 2011].

USGCRP (U.S. Global Climate Change Research Program). 2009. Global Climate Change Impacts in the United States. A State of Knowledge Report from the U.S. Global Change Research Program [online]. Available: http://www.global change.gov/publications/reports/scientific-assessments/us-impacts/download-the-report [accessed Dec. 27, 2011].

USEtox.org. 2009. Background of the U.S. Etox^{TM} model [online]. Available: http://www.usetox.org/background.aspx [accessed Mar. 7, 2011].

Wang, S., J. Hao, M.S. Ho, J. Li, and Y. Lu. 2005. Intake fractions of industrial air pollutants in China: Estimation and application. Sci. Total Environ. 354(2-3):127-141.
WHO (World Health Organization). 2012. Environmental Health Inequalities in Europe: Assessment Report. World Health Organization [online]. Available: http://www.euro.who.int/__data/assets/pdf_file/0010/157969/e96194.pdf [accessed May 16, 2012].
Wild, C. 2005. Complementing a genome with an "exposome": The outstanding challenge of environmental exposure measurement in molecular epidemiology. Cancer Epidemiol. Biomarkers Prev. 14(8):1847-1850.
Wild, C.P. 2012. The exposome: From concept to utility. Int. J. Epidemiol. 41(1):24-32.
Willett, W.C. 2002. Balancing the life-style and genomics research for disease prevention. Science 296(5568):695-698.
Woodruff, T.J., A.R. Zota, J.M. Schwartz. 2011. Environmental chemicals in pregnant women in the United States: NHANES 2000-2004. Environ. Health Perspect. 119(6):878-885.
World Bank. 2009. World Development Report 2009: Reshaping Economic Geography. Washington, DC: World Bank [online]. Available: http://www-wds.worldbank.org/external/default/WDSContentServer/IW3P/IB/2008/12/03/000333038_20081203234958/Rendered/PDF/437380REVISED01BLIC1097808213760720.pdf [accessed May 29, 2012].
Ziegier, R.G., R.N. Hoover, M.C. Pike, A. Hildesheim, A.M. Nomura, D.W. West, A.H. Wu-Williams, L.N. Kolonel, P.L. Horn-Ross, J.F. Rosenthal, and M.B. Hyer. 1993. Migration patterns and breast cancer risk in Asian-American Women. J. Natl. Cancer Inst. 85(22):1819-1827.

5

Scientific and Technologic Advances

INTRODUCTION

Obtaining reliable estimates of exposures of large populations on multiple scales of space and time requires detailed information on emissions or transformation products from a source, on the locations of receptors (personal or ecosystem) in time and space, and on the activity levels of the receptors (as proxies for inhalation rate, ingestion potential, or dermal interaction) at the time when they are affected by a source. Additional information may be needed on how individual heterogeneity influences exposures—including such susceptibility characteristics as genetics, pre-existing health conditions, and psychosocial stress—because these factors may also influence exposure. Information on body burden, obtained by collecting exposure biomarkers, also is essential for understanding the dose from a specific source and the influence of environmental exposures on health risks.

Efforts to characterize exposure have focused on ambient conditions, and an individual is typically assigned to a home address in an epidemiologic health study or a species is assigned to a region it inhabits in an ecosystem. Although those exposure assignments have revealed important health risks, reliance on proxy methods may impart large exposure-measurement error—that is, a modeled exposure may be an inaccurate and potentially biased estimate of the true exposure. Depending on the exposure-error type, health-effect estimates may be attenuated and biased toward a null result, and the true benefits of control measures may be obscured. Obtaining more accurate estimates of internal exposure reduces exposure-measurement error and provides a more realistic understanding of potential health effects of environmental and occupational exposures (Carroll et al. 2006).

Efforts to gather information on personal exposures have relied on specialized equipment that is expensive and cumbersome and thus limits the wear time or number of subjects that can be monitored. Because of those limitations, many studies have used questionnaires (Wacholder et al. 1992) or simple information on location, such as home address, that is related to exposures. Those techniques have well-known limitations, but they are often the only methods available, particularly for reconstructing historical exposures.

Innovations in science and technology provide opportunities to overcome limitations and guide exposure science in the 21st century to deliver knowledge that is effective, timely, and relevant to current and emerging environmental-health challenges. Personalized medicine[1] and telemedicine will increase the pace of innovation in scientific and technologic methods that will benefit the field of exposure science. For example, many new genomic methods for monitoring individual metabolic and exposure phenotypes will be critical for future individualized medicine. In telemedicine, cellular-telephone technologies increasingly contribute to improving diagnostics and patient care and hence to improving our ability to anticipate the effects of exposures (Wootton and Bonnardot 2010). Similarly, new developments in geographic information science and technologies are leading to rapid adoption of new information obtained from satellites via remote sensing, which provides immediate access to data on potential environmental threats. Improved information on physical activity and locations of humans and other species obtained with global positioning systems (GPS) and related geolocation technologies is increasingly being combined with cellular-telephone technologies. Many of these advances are integrated through powerful geographic information systems (GIS)[2] that operate either through stand-alone computing platforms or through the World Wide Web. Biologic monitoring and sensing increasingly offer the potential to assess internal exposures. The convergence of these scientific methods and technologies raises the possibility that in the near future embedded, ubiquitous, and participatory sensing systems will facilitate individual-level exposure assessments on large populations of humans or other species.

The new technologies and methods also may help to operationalize the concept of the exposome (see discussion in Chapter 1). Establishing a more complete record of exposures based on internal biomarkers as theorized in the exposome (Wild 2005) requires tools that can also assess external environmental exposures. Many important exposures lead to no internal biomarkers but can be associated with environmental health risks (Peters et al. 2012) (for example, noise, heat, and electromagnetic fields). There is also a need to continue to link sources to exposures; this is the basis of mitigation efforts to protect public health. The committee envisions that many of the new technologies discussed in this chapter will help to broaden the exposome to the "eco-exposome" concept discussed in Chapter 2, and help to quantify exposure indicators to address those concerns.

---

[1]Personalized medicine is an emerging practice of medicine that uses information about a person's genetic profile and environmental exposures to prevent, diagnose, and treat disease (Offit 2011).

[2]GIS is defined as a system for performing numerous operations involving the acquisition, editing, analysis, storage, and visualization of geographic data (Longley et al. 2005).

In this chapter, the committee reviews some of the newest technologies for exposure science and, in considering their strengths and limitations, identifies near-term and long-term innovations that will guide exposure science in the 21st century. The review is organized according to the framework in Chapter 1 that describes the scope of exposure science from characterizing external concentrations to personal exposures and finally to understanding how internal exposures affect dose. Figure 5-1 expands on the framework by identifying the technologies that will be presented. The discussion begins with a review of geographic information technologies, which help in characterizing sources and concentrations and also can improve understanding of stressors and receptors when used in concert with other methods and information. Ubiquitous sensing systems, ecologic momentary assessment (participatory methods that are used to query subjects about their perceptions and experiences while in the exposure field using cell phones or other real-time devices), and nanosensors are addressed next; these can help in characterizing personal exposures. We then discuss biomonitoring, which can improve our understanding of internal exposures and, when combined with other technologies, can help to identify sources. Finally, models and information-management tools are addressed in the context of their ability to help in interpreting and managing the massive and often complex interactions among receptors and environmental stressors. Many of the technologies in this chapter are illustrated in connection with air pollution, inasmuch as this is one of the most developed sectors of exposure science. As shown in Figure 5-1, however, the committee's framework and vision are intended to be broadly applicable and relevant to all media to reflect the expected needs for the technologies, and many other illustrative examples are presented.

TRACKING SOURCES, CONCENTRATIONS, AND RECEPTORS WITH GEOGRAPHIC INFORMATION TECHNOLOGIES

Three major technologic advances in geographic information technologies—remote sensing, global positioning and related locational technologies, and GIS—have dramatically affected exposure science. As outlined by Goodchild (2007), they are inspiring a new emphasis on spatial information in relation to social and scientific inquiry. Over the last 10 years, the technologies have contributed to improvements in exposure science, and they will probably continue to move the field toward more refined exposure assessments that are more comprehensive, more accurate, and more relevant to and valuable in policy-making and in the everyday lives of large populations.

Remote Sensing for Exposure Assessment

Remote sensing (RS) has emerged as a key innovation in exposure science. RS has been defined as "the acquisition and measurement of data/information on some property(ies) of a phenomenon, object, or material by a recording

Scientific and Technologic Advances 109

device not in physical, intimate contact with the feature(s) under surveillance" (Short 2011). The field encompasses the capture, retrieval, analysis, and display of information on subsurface, surface, and atmospheric conditions that is collected with satellite, aircraft, or other technologies designed to sense energy, light, or optical properties at a distance (Jerrett et al. 2009). RS is an important tool for enhancing the capacity to assess human and ecologic exposures because it provides global information on the earth's surface, water, and atmosphere. It is also widely used for subsurface investigations (for example, electromagnetic imaging of karst in water resource investigations). It also provides exposure estimates in regions where sparse ground observation systems are available.

With respect to air pollution, the most common aerosol characteristic measured with a satellite is the aerosol optical depth (AOD), which quantifies the extinction of electromagnetic radiation from aerosols in an atmospheric column at a given wavelength (Emilli et al. 2010). Six primary satellite sensors provide information on particulate pollution (MODIS, Landsat, IKONOS, Orbview, SPOT, and GOES). Box 5-1 discusses evaluation of the reliability of AOD compared with $PM_{2.5}$[3] mass concentrations measured on the ground. Box 5-2 and Figure 5-2 demonstrate how the results of a 1-km retrieval of the MODIS AOD substantially improve the resolution and thus the utility of remote sensing for health and ecologic studies; the current grid size has a 10-km retrieval.

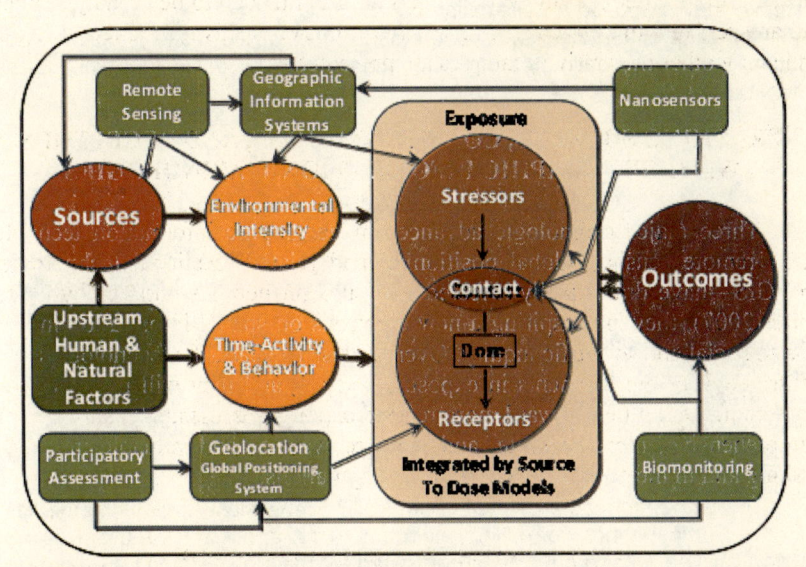

FIGURE 5-1 Selected scientific and technologic advances considered in relation to the conceptual framework.

---

[3]$PM_{2.5}$ are fine particles in the ambient air that are 2.5 microns or less in diameter.

> **BOX 5-1** Evaluating the Reliability of Aerosol Optical Depth Against Ground Observations
>
> Hoff and Christopher (2009) reviewed more than 30 papers that examined the relationship between total column AOD and surface $PM_{2.5}$ measurements on a station by station basis. Their results underscored "the range of measurements from across the globe and the range of correlations between AOD and mass". They found a wide range of uncertainty between the two measures of AOD and mass. The studies used simple linear regressions and correlations between the AOD values and the $PM_{2.5}$ mass concentrations measured on the ground. In some cases the correlations were strong and the AOD served as a predictor of pollution on the ground. In other cases, either because the satellite product itself was not sufficiently accurate or because the particles observed in the total column were in layers aloft, the satellite derived AOD was a poor predictor of pollution at the earth's surface. The authors suggested conducting a study of the controlling extrinsic factors for each region that would aid in understanding the $PM_{2.5}$–AOD relationship. The literature continues to grow with efforts to combine information from multiple satellite sensors and models (van Donkelaar et al. 2010) or to introduce auxiliary information, such as meteorologic data (Pelletier et al. 2007) or boundary layer height (Engel Cox et al. 2006). Lee et al. (2011) have hypothesized that the inherent variability in the PM–AOD relationship is due to changes in particle size and composition, earth surface properties, vertical distribution of particle concentrations, and other factors. To account for the variability of these factors, they proposed a daily calibration technique that is based on the spatial variability of ground PM measurements and would make it possible to obtain quantitative estimates of PM concentrations by using AOD measurements.

> **BOX 5-2** Evaluation of MODIS 1 km Product
>
> The development of the 3 km and 1 km products provides an opportunity to test the capabilities of the satellite data to provide the resolution needed for exposure assessment and health related studies. For example, the 10 km aerosol product offered by MODIS is sufficient for climate applications but insufficient for detailed exposure assessment from sources that are variable over small areas, such as traffic emissions. In that regard, Hoff and Christopher (2009) stressed the importance of a finer resolution product on a local urban scale. It is expected that a 3 km product will become publicly available in 2013.
>
> To attain 1km resolution AOD from MODIS, the Multi Angle Implementation of Atmospheric Correction algorithm was applied (Lyapustin et al. 2011a,b). The 1 km product was generated for the New England area during 2003. Figure 5-2 compares the 10 km and 1 km retrievals. It clearly shows that considerably more detail is obtained with the 1 km product.

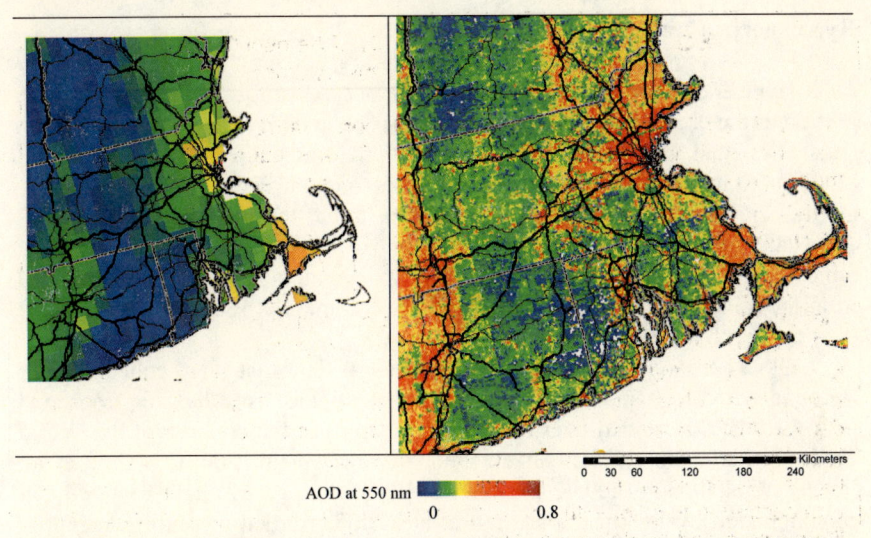

FIGURE 5-2 Aerosol optical depth (AOD) derived from MODIS data for the New England region with the standard 10-km algorithm (left) and the experimental 1-km algorithm (right) for June 25, 2003.

Remote Sensing for Health, Exposure, and Ecologic Studies

Several studies and reviews (for example, Maxwell et al. 2010) have suggested that higher-resolution data enhance efforts to identify time–space patterns that are the basis of many risk assessments for diseases (Wilson 2002). In many studies, remote sensing data were used to derive three variables: vegetation cover, landscape structure, and water bodies. The ability to sense vegetation remotely from space is important in that nearly all vectorborne diseases are linked to the vegetative environment during their transmission cycle. Furthermore, crop-type information may be important for studying the effects of pesticides (for example, vector resistance and illnesses caused by exposure to toxins) (Beck et al. 2000). Ward et al. (2000, 2006) and Maxwell et al. (2010) used crop location to identify where pesticides were applied in relation to residential locations (Maxwell et al. 2010). Remote sensing of vegetation cover combined with GIS has also been used to develop management strategies to reduce herbicide application (Gómez-Casero et al. 2010) and to assess potential exposure of fish and wildlife to pesticides and metals (Focardi et al. 2006).

Green cover is also associated with higher levels of physical activity, and RS has been used with geolocation technologies to show associations between physical activity of children and their exposure to green cover (Almanza et al. 2012).

Hyperspectral Imaging

Hyperspectral imaging collects and processes information from a wide portion of the electromagnetic spectrum. It has been used to assess human health risks associated with infectious diseases or environmental hazards. Ong et al. (2003) used hyperspectral airborne techniques to quantify dust loadings on mangroves originating from mining. The authors found that they could detect and quantitatively map the distribution of iron oxide. Ferrier (1999) showed that mine tailings (which contain potentially toxic materials) had been dispersed from the mine workings extending down the Rambla del Playazo to within 600 m of the beach at El Playazo (Spain).

Other researchers have used hyperspectral data that were collected over "Ground Zero" for rapid assessment of the potential asbestos hazards associated with the dust that settled over lower Manhattan after the collapse of the World Trade Center towers (Clark et al. 2001; Swayze et al. 2006). Malley et al. (1999), Winkelmann (2005), and van der Meer et al. (2002) have reported soil contamination by hydrocarbons. Wu et al. (2005) studied mercury contamination in suburban agricultural soils in the Nansing region of China. Finally, Chudnovsky et al. (2009, 2011) used the Hyperion satellite data to separate the spectral features of the Saharan dust storm from the underlying surface.

HI sensing has been used to examine exposures of coral reefs to stressors such as sea surface temperature, ultraviolet radiation, wind, sediment load, chlorophyll, acidification, salinity, and coastal development (Maina et al. 2008; Eakin et al. 2010). Sediment load/water clarity (Doran et al. 2011), stressor exposures in benthic ecosystems (Goetz et al. 2008), and other water quality parameters (Bagheri and Yu 2008; Odermatt et al. 2012) have also been analyzed using HI techniques. More recently HI techniques were utilized to assess the extent of the Deepwater Horizon Oil Spill and possible exposures to oil in pelagic and nearshore ecosystems (Bradley et al. 2011; Lavrova and Kostianoy 2011; Bulgarelli and Djavidnia 2012; Mishra et al. 2012).

In 2015, two new hyperspectral sensors will be launched: the National Aeronautics and Space Administration HyspIRI (NASA 2011) and European ENMAP (EnMAP 2011) missions. With their improved hyperspectral and multispectral capabilities, these sensors will increase the ability to monitor the effects of urbanization on the environment and to assess land-cover characteristics that could indicate the presence of or risks posed by vectorborne and animal-borne diseases on a global scale.

Conclusions

To improve data quality for RS and increase its utility for exposure studies, technologic improvements are needed, including

- Breakthroughs in electro-optics technologies.

- Improvement of the current AOD retrievals (to achieve near-laboratory air-quality data) by obtaining accurate and reliable atmospheric vertical profile information.
- Retrieval of high-resolution AOD to discern spatial patterns of pollution in urban environments through frequent daily temporal coverage based on orbital sensors.

Global Positioning System and Geolocation Technologies

Launched in the 1980s for defense applications, the GPS offers exposure scientists a simple means of tracking the geographic position of a person or other species. GPS receivers are now embedded into many cellular telephones, vehicle navigation systems, and many other instruments (Goodchild 2007). The GPS is a utility owned by the US government, and it consists of three components: a space segment with at least 24 satellites that transmit one-way signals to the earth; a control segment that maintains ground stations to track the satellites, reset their clocks, and maintain their positions; and a user segment that consists of individual devices that users deploy to receive the signals and calculate three-dimensional positions and times (GPS 2011). GPS signals can be augmented or complemented by land-based navigation systems that use cellular-telephone triangulation to provide positions when satellite signals are unavailable because of, for example, topographic obstruction or weather conditions (Shoval and Isaacson 2006). Radiofrequency identification can also be used for local tracking of goods, animals, or people (Goodchild 2007). Collectively, these systems are referred to here as geolocation technologies.

Hundreds of studies have used geolocation technologies to improve assessment of environmental exposures, including exposure to infectious-disease vectors (Vazquez-Prokopec et al. 2009) and air pollution (Paulos et al. 2007); to analyze how physical activity is related to different built environments (Jones et al. 2009); and to inform simulation models of potential pesticide exposures (Leyk et al. 2009). Many other applications are found in the literature.

The main contribution of geolocation technologies is to reduce exposure measurement error and to move closer to a "time–geography of exposure" (Hagerstrand 1970; Briggs 2005). That is, geolocation technologies offer the possibility to know, with a high degree of accuracy, an individual's location in time and space and to provide a window into the moment of contact between a source (that is, an environmental intensity) and a receptor. When data obtained on environmental intensities (for example, air or water quality) are combined with geolocation information and physical activity measurements (obtained with accelerometers), more detailed estimates of potential chemical, biologic, or physical exposures can be made by using data on inhalation rate, ingestion potential, or dermal contact.

There are many examples of how geolocation technologies have improved our understanding of exposures through their use in defining a person's location in time and space. They have revealed important limitations of survey-based assessments of location. For example, a study by Elgethun et al. (2007) compared time–activity diaries with actual measurements from GPS and found severe underreporting in the diaries regarding the amount of time spent outdoors at home. Such errors may result in substantial exposure misclassification to such pollutants as ozone that have low penetration ratios from outdoors to indoors, which make time outdoors a key determinant of exposure. The technologies provide more accurate information on geocoded locations of subjects and a better understanding of likely sources of error when points are used to represent large structures, such as schools and day-care facilities (Houston et al. 2006). They are also used in studies in which exposures are measured as study subjects walk, ride bicycles, or drive with pollution monitors. A study by McCreanor et al. (2007) demonstrated the effect of walking through polluted areas on asthmatic symptoms and biomarkers—such as exhaled nitric oxide, a marker of lung inflammation—in London, England. The study provided increased support for the hypothesis that ambient air-pollution concentrations can elicit changes in asthmatic symptoms. Geolocation technologies have already made important contributions to the understanding of exposures at the point of contact between source and receptor, and they appear poised to play an increasingly integral role in widespread population-based individual sensing (discussed below).

Geographic Information Systems

GIS combines topologic geometry, capable of manipulating geographic information, with automated cartography and enables users to compile digital or hard-copy maps. GIS plays a central role in integrating data into coherent databases that connect different attribute data (for example, exposure and health attributes) by geographic location. Input data used to derive exposure surfaces, such as road locations and industrial land uses, also are stored and manipulated in GIS. GIS increasingly serves as the storage and integrative backbone of remote sensing, geolocation technologies, and sophisticated modeling outputs, such as for collecting measurements on the fate and transport of contaminants through ecosystems (Gallagher et al. 2010).

Another important role of GIS in exposure assessment is the quantification of topologic relationships. For example, buffer functions that measure the distance between a source, such as a roadway, and a receptor, such as a house, enable analysts to relate the geographic position of a study subject in space and time with the subject's likely exposure on the basis of an overlay of location information (Jerrett et al. 2005). That type of buffering, which provides the distance between a source and a receptor, is used to characterize proximity to roadways, factories, water bodies, and other land uses or modifications that have either potentially adverse exposures (for example, pesticide transport from agricultural fields) (Gunier et al. 2011) or potentially favorable exposures (for example, parks and health-food stores in cities) (Morland and Evenson 2009). GIS can provide information that is stored by the user, both before and after a major change (for example, in land use) or a catastrophe (for example, a tsunami). Figure 5-3 demonstrates a road buffer that was used to characterize human exposures to traffic-related air pollution in Hamilton, Canada (Jerrett et al. 2005). Ecologic studies have combined modeling results with overlay techniques to examine potential exposure exceedances of threatened and endangered species (for example, see Figure 5-4 for cadmium exposures of the Little Owl) (Lahr and Kooistra 2010).

Web-Based Geographic Information Systems for Exposure Assessment

Web-based GIS is becoming more common (Maclachlan et al. 2007) and can serve as a tool in policy-making and in educating and empowering communities to understand and manage their environmental exposures better. (See Chapter 6 for additional discussion of community engagement.) For example, to promote active commuting, Metro Vancouver has collaborated with the University of British Columbia to develop a cycling-route planner (Cycling Metro Vancouver 2007), which allows cyclists to select routes that have the most green vegetation, the least traffic pollution, and the least or greatest elevation, all specified by the user. That empowers cyclists to choose the routes that best suit their fitness levels, minimize exposure to traffic pollution, and reduce their carbon dioxide output (Su et al. 2010). The Web site runs on the backdrop of Google Maps—an illustration of the potential synergies between new private-sector technologies and public-health protection.

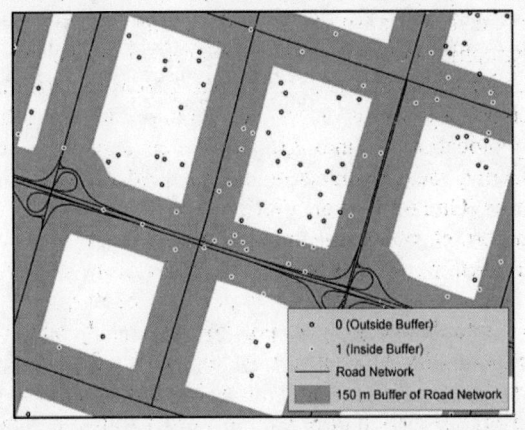

FIGURE 5-3 Example of a binary buffer overlay showing people likely to experience traffic-related air-pollution exposure. The circles represent people. People assigned a "0" are outside a prespecified distance, while people assigned a "1" are within a given distance. Adapted from Jerrett et al. 2005. Reprinted with permission; copyright *2005, Journal of Exposure Science and Environmental Epidemiology*.

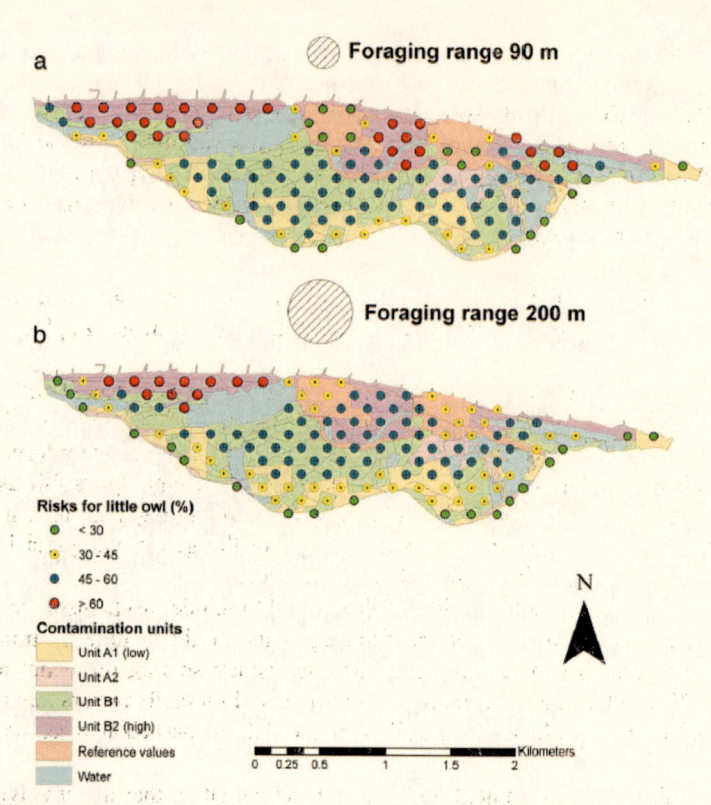

FIGURE 5-4 Map of a flood plain in the Netherlands showing secondary risk of poisoning by cadmium in Little Owls developed using a combination of measured cadmium concentrations, food web modeling, knowledge of foraging in different habitats, and probabilistic risk assessment. Source: Lahr and Kooistra 2010. Reprinted with permission; copyright 2010, *Science of the Total Environment*.

In addition to human health concerns, web-based GIS has been used to monitor ecologic exposures. For example, Google Earth and Google Fusion Tables with Airborne/Visible Infared Imaging Spectrometer data (AVIRIS 2012) were used to provide public, real mapping of the Deepwater Horizon Oil Spill (Bradley et al. 2011).

With the new technologies—such as cellular telephones, GPS, and computers that apply complex data-mining techniques—private companies are increasingly collecting data that are potentially useful for exposure science, such as location and mobility information, and in some cases direct measurements of exposure through sensing networks. Issues of data ownership, use, informed consent, and data-sharing remain to be addressed. Increased cooperation with private-sector entities offers great potential for enhancing the data available for exposure science.

Conclusions

GPS and GIS have already contributed in important ways to exposure assessment, enabling researchers to refine assessments and to understand how to move from assessing ambient concentrations to understanding the likely exposures received by people. The committee suggests the following measures to capitalize further on these advances:

- Continued support for the GIS "infostructure" to ensure that public-health agencies and researchers have access to the wealth of geographic exposure data.
- Expanded access to existing exposure data that are collected with public funding for researchers conducting scientific research. Many of the existing data sources permit only partial release of critical information or the information is only released in highly restricted data enclaves, often limiting its full use by scientists. Agencies responsible for collecting and maintaining these data should make all efforts to expand access, both in terms of the geographic precision of the data and the availability of the data to researchers.
- Efforts to promote sharing of data should include mechanisms to validate the data collected with federal funding. This would include a minimum adherence to established requirements for interagency sharing of spatial metadata (FGDC 2011). These metadata for sharing should include factors that allow a potential user to evaluate applicability of the data to particular research projects.
- Efforts by government agencies and universities that are involved in exposure-science research should work to foster cooperation with the private sector to encourage data collection, sharing of geographic and exposure information, and the formation of partnerships with exposure scientists with the goal of improving public-health protection.
- Efforts to support and enhance Web-based exposure mapping (for example, Cycling Metro Vancouver 2007) to improve access to data on and understanding of potential exposures.

UBIQUITOUS SENSING FOR INDIVIDUAL AND ECOLOGIC EXPOSURE ASSESSMENT

Limitations of Current Environmental and Personal Monitors and the Potential of Ubiquitous Sensing

Lack of personal and individual exposure data is one of the greatest limitations in exposure science. A challenge in addressing the paucity of data is imputing, interpolating, or modeling the likely exposures that are not directly measured on a person or on a species in its ecosystem. The void in fine-scale exposure data on an individual or a species means that considerable error may be introduced in assessing dose–response relationships. Assessing the influence of that measurement error on dose–response functions is challenging because researchers typically lack sufficient information on the "gold standard"—measured exposure information on the individual or in key microenvironments. As a result, exposure assessment has been called the Achilles heel of environmental epidemiology (Steenland and Savitz 1997).

The last 20 years have seen substantial technologic advances in personal environmental monitoring. Despite the advances, however, personal sensors are still inadequate in their capacity to obtain highly selective, multistressor measurements in out-of-the-laboratory environments. (Measurements of multiple stressors provide an understanding of exposure to mixtures that may elucidate our understanding of health responses.) Many personal monitors rely on pumps to collect air samples on filters or other sorbent materials, which are then analyzed to measure integrated average concentrations over time (usually days or weeks). However, the size, weight, noise level, and appearance of monitors are unacceptable to some users, and this shortens monitoring times, lowers compliance, and introduces bias. Valuable information on the locations and conditions (for example, sources) where and when exposures occur and on the frequency and magnitude of transient high-level exposures is not captured in longer-term integrated samples. In general, current techniques do not consider the physical activity of the subject during exposure. Passive samplers (for example, badges or sampling tubes) make it possible to collect samples without pumps but are limited in the analytes that can be measured at environmental concentrations and often require longer sampling periods (for example, 1-2 weeks) to obtain sufficient concentrations in nonoccupational environments. The analytic costs of each sample may be high.

Innovations in personalized monitoring and in analysis of a person's mobility can ameliorate those shortcomings—in particular, through sensing of the environment through modification of cellular telephones, which are carried routinely by billions of people around the world. Many cellular telephones come equipped with motion, audio, visual, and location sensors, and several software applications have been written to exploit the on-board sensors through cellular or wireless networks (see, for example, Seto et al. 2010). Other devices, such as pollution monitors, can be either built into the telephones or connected through a body-sensing network via Bluetooth radio.

Cellular telephones, supporting software, and the expanding networks (cellular and WiFi) potentially can be used to form "ubiquitous" sensing networks to collect personal exposure information on millions of individuals and large ecosystems using citizen–scientists (Crall et al. 2010). Such networks can also take advantage of "embedded" sensors installed in existing infrastructure, for example, weight-in-motion sensors installed on roadways or sensors installed in public vehicles, such as buses, that provide anonymous, continuous data collection. In addition, there is participatory sensing, which differs from other embedded networks in that the people in the network knowingly and voluntarily collect information on environmental conditions (or their own mobility) in exchange for some individual benefit, such as increased knowledge about their

Scientific and Technologic Advances 119

own environmental exposures or information on their level of physical activity (Burke et al. 2006). All these research approaches, however, pose challenges of bias and compliance. Boxes 5-3, 5-4, and 5-5 provide examples of embedded, ubiquitous, and participatory sensing, respectively; these categories, however, are not mutually exclusive, in that they may include elements of one another.

BOX 5-3 Embedded Sensing of Traffic in Rome

Embedded sensors are now being piloted in the city of Rome, Italy to help with tracking mobility patterns of pedestrians, bicyclists, and vehicles and for managing traffic flows. Given the role of traffic in air pollution, noise, and accident risk, better information on traffic and other modes of transportation are needed. Rome's embedded sensor system relies on a Telecom Italia's Localizing and Handling Network Event Systems (LocHNESs) software platform, which uses anonymous information on location from cell phone users in combination with embedded location tracking from public transit vehicles. The system is being tested to supply near real-time traffic monitoring and management information (Calabrese et al. 2011). Such information on traffic could be combined with models to estimate noise or air pollution levels throughout the city. The system can output data into a variety of formats, including a 40*40 m grid cell resolution showing levels of traffic congestion.

LocHNESs illustrates how ubiquitous mobile phones can supply anonymous information on location that can be combined with embedded tracking networks on public infrastructure to deliver real-time data on environmental exposures.

BOX 5-4 Ubiquitous Sensing of Physical Activity and Location

Increasing availability of ubiquitous mobile devices—particularly smart telephones with motion sensing, GPS, and wireless capabilities—has created opportunities to develop new tools and methods to study and intervene to address sedentary lifestyles, obesity, and ambient risk factors, such as air pollution, noise, or ultraviolet radiation. One example of the potential is CalFit software. CalFit is an application that runs on mobile telephones that use the Android operating system. The software uses the accelerometry and GPS sensors that are typically built into all smart telephones to record activity counts and energy expenditure and the time and location in which an activity occurs. The device consists of a single telephone that can be carried and used as a normal telephone by participants in a study or by the general public. The software program has a single on–off switch for data-logging and, once turned on, will continuously collect data as a background service and not stop until turned off (Seto et al. 2010). Pilot studies testing CalFit with 35

(Continued)

BOX 5-4 Continued

free-living human volunteers in Barcelona, Spain, indicate that the software collects data on location and physical activity that compares well with commercially available stand-alone triaxial accelerometers (de Nazelle et al. 2011; Seto et al. 2011) (see Figure 5-5). When combined with dispersion or other models of ambient exposure, CalFit offers the potential for adjusting, for example, the dose of air pollution received by a person to account for exposure to air pollutants, rather than relying on exposure concentration in the home or workplace. The system is also capable of serving as a base station for other sensors that operate via Bluetooth radio to collect such data as air pollution, light, and noise.

CalFit and other cellular-telephone–based systems can also be used to implement context-specific ecologic momentary assessment (CS-EMA). CS-EMA measures real-time exposures and outcomes with sensors that are inside and outside the telephone. The system can communicate information, telling a person to respond to a survey when particular events are observed, such as a period of physical activity, exposure to air pollution, use of steroid inhalants, or consumption of particular food. Responding to these surveys provides opportunities to obtain important information about an exposure or outcome, such as mood, stress, behaviors, and other information (Intille 2007; Duntun et al. 2010).

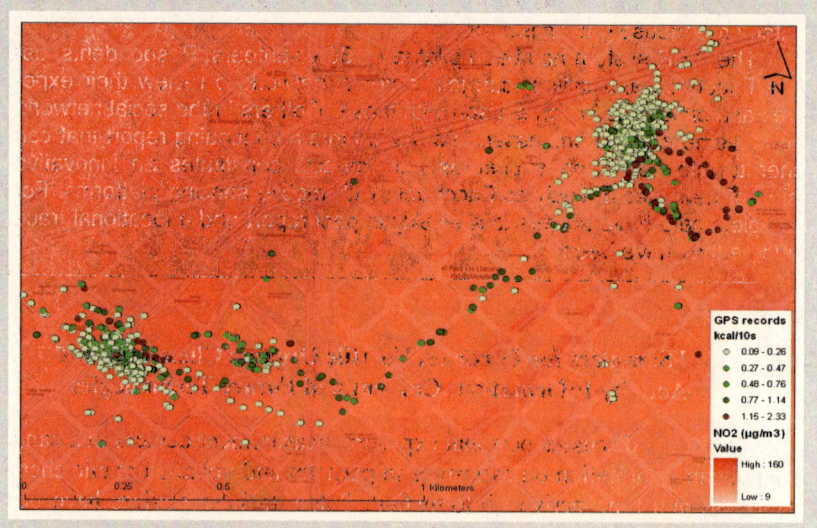

FIGURE 5-5 Output from a CalFit telephone showing the location and activity level of volunteers in kilocalories per 10-second period in a pilot study in Barcelona, Spain. Activity traces are overlaid onto a map of nitrogen dioxide concentrations. Source: Map supplied courtesy of de Nazelle et al. 2011. Reprinted with permission from the author; copyright 2011.

BOX 5-5 Participatory Sensing

Participatory sensing refers to systems of distributed data collection and analysis in which participants decide on what, where, and when to monitor in their environments (Mun et al. 2009). Such systems operate on various scales—for example, individual, group, urban, and global—depending on what is being measured (Burke et al. 2006). Participatory sensing systems often combine embedded and ubiquitous systems with Web-based applications that allow participants to share information on their exposures and to understand exposures of others in the participatory system.

One example is the Personal Environmental Impact Report (PEIR) system that operates in Los Angeles, California (Mun et al. 2009). The system measures four main outcomes: exposure to fine particulate matter ($PM_{2.5}$), exposure to fast food outlets, output of transportation-related greenhouse-gas emissions, and output of transportation-related $PM_{2.5}$ emissions near sensitive receptors. PEIR relies on cellular-telephone locational and speed information and on a sophisticated activity-classification system that uses information from GPS, cellular-telephone towers, and data on land use and traffic. Activities are classified as walking or driving with a Markov chain algorithm.

Once activities are classified, exposures can be assessed through a near real-time dispersion model for $PM_{2.5}$ that combines the likely exposure levels and a person's location to assign a likely concentration. People also can examine their exposure to fast food, generation of transportation-related $PM_{2.5}$ emissions near sensitive receptors, or their impact on transportation-related greenhouse gas emissions.

The PEIR system has been piloted by 30 volunteers. Respondents, using a Facebook application, can join a social network to review their exposures and emissions in comparison with those of others in the social network. The movement of the processed information into a continuing report that can generate information on exposures and impacts constitutes an innovative fusion of new media, such as Facebook, with mobile sensing platforms. For example, users have access to a weekly impact report and a locational trace of where they have been.

Personal Samplers for Particles, Volatile Organic Chemicals, and Time–Activity Information: Current and Future Technologies

This section discusses personal exposure measurement devices that can be used to obtain information on exposures to particles and volatile organic chemicals (VOCs) and time–activity information. Some of these devices have been fully tested; others are new and yet to be implemented.

Current Microsensor Technologies

This section addresses state-of-the-art microsensor systems. Some systems are currently in the prototype stage, while others are commercially available.

Microsensor Systems

Miniature microsensor systems incorporating preconcentrator, microfabricated gas chromatography (micro GC) systems, and microsensor arrays have been developed for detecting VOCs with detection limits below parts per billion (Kim et al. 2011). The sensor systems include a miniature pump for sample collection. The VOCs are first separated by a high-performance GC system before detection by sensor arrays (Kim et al. 2011; Kim et al. 2012a,b). Sandia National Laboratory has also developed a microsensor system based on micro GC and surface acoustic-wave sensors (Lewis et al. 2006).

Volatile-Organic-Chemical Monitor (Arizona State University HYBRID)

A portable ("wearable") VOC monitor was recently developed (Iglesias et al. 2009; Chen 2011). It uses a "hybrid chemical sensor" for nearly real-time measurement and wireless transmission of data via a cellular telephone. The sensor combines a miniature gas chromatographic system for preconcentration and separation of chemical species with a novel detector that consists of a tuning fork coated with molecular imprinted polymers. The operation is controlled by an internal microprocessor. Benzene, toluene, ethylbenzene, and xylene components can be individually measured at concentrations as low as 1 ppb. The components and operating characteristics can be varied to measure other types of VOCs.

Pretoddler Inhalable Particulate Environmental Robotic Bioaerosol Sampler

A Pretoddler Inhalable Particulate Environmental Robotic (PIPER) personal sampler to measure indoor particulate matter (PM) and bioaerosols was recently developed (Shalat et al 2011; Wang et al. 2012). It is used as a surrogate to substitute for direct personal monitoring of very young children (it is not practical for them to wear conventional personal samplers). The PIPER sampler can hold up to two personal air-sampling devices and can mimic the speed and pattern of motion and the breathing height of boys and girls in three age groups: 6 months–1 year, 1–2 years, and 2–3 years.

Dual-Chamber Particle Monitor

A small, portable, data-logging particle monitor was recently developed (Litton et al. 2004; Edwards et al. 2006; Chowdhury et al. 2007). The device combines an ionization detector that is sensitive to submicrometer particles with a photoelectric detector that is sensitive to micrometer particles. Because the detection limit is 50 $\mu g/m^3$, it is not sensitive enough for typical ambient concentrations of particles found in many developed countries. However, the monitor was designed for use in locations with much higher concentrations of particles, such as households in many developing countries, including kitchens with wood stoves. Simplicity of operation, low cost, battery operation, low weight and small size, and quiet operation are potentially useful features.

University of California, Berkeley Time–Activity Monitoring System

A time–activity monitoring system was recently developed by Allen-Piccolo et al. (2009). It uses small, lightweight ultrasound transmitter worn by participants and an ultrasound receiver (locator) attached to a data logger that is fixed at indoor locations to record accurately when and how long each participant is in each location. The method provides more reliable time–activity information than is typically available in diaries maintained by participants. The time–activity information is used to estimate overall personal exposures on the basis of exposure measurements made with fixed monitors in each microenvironment.

Black-Carbon Monitor (microAeth)

This instrument is small, lightweight, portable, and battery-operated and is commercially available (AethLabs 2012). It measures real-time black carbon by using light transmittance for particles collected on Teflon-coated glass-fiber filters. Its dimensions are 4.6 x 2.6 x 1.5 inches, and it weighs 0.62 lb. Flow may be 50, 100, or 150 mL/min. Battery power will operate the instrument for a minimum of 24 hours with a flow of 100 mL/min and a 5-minute time base.

Personal Multipollutant Sampler

This compact personal and microenvironmental monitoring system, developed by Chang et al. (1999) and Demokritou et al. (2001), allows simultaneous active collection of particles on filters and passive sampling of pollutant gases. Different configurations allow use of selected combinations of Teflon membrane filters to measure PM_{10} and $PM_{2.5}$ mass, $PM_{2.5}$ trace elements with x-ray fluorescence and inductively coupled plasma mass spectrometry, black carbon with reflectance, elemental and organic carbon (on quartz fiber filters) with thermal optical methods, and passive ozone, nitrogen dioxide, and sulfur dioxide with diffusion badges. The sampler can be mounted on a backpack strap on the subject's chest, and the battery-operated pump is carried in the backpack (it allows 24 hours of continuous sample collection).

RTI Micro-miniature Personal Exposure Monitoring

A personal particle monitor recently developed by RTI International, microPEM Model v3.2 (RTI 2008), is a small, lightweight nephelometer (using light-scattering) that provides nearly real-time particle concentration data (PM_{10} or $PM_{2.5}$) with simultaneous collection of particles on a gravimetric filter. With careful filter measurements and 24-hour sampling, the monitor is suitable for particle concentrations as low as 3.6 µg/m^3.

Future Developments with Nanosensors

Developing ubiquitous monitoring networks for personal exposure assessment will depend on rapid advances in pollution-sensor technologies. Although advances have been made with the technologies described, most of them are still not capable of measuring multiple pollutants continuously. A technology needs assessment conducted by the Oak Ridge National Laboratory determined that there was a need for a rugged, lightweight, low-cost, wearable, real-time sensor capable of multianalyte detection with minimal burden on the person using the monitor (Sanchez et al. 2010). Such a sensor would need to be able to detect acute and subacute chemical agents simultaneously with the same sensing system used in the field and then link the data to a specific biologic event. The device would need to be capable of remote data acquisition, location recording, and measurement of both the level and frequency of the environmental exposure (Sanchez et al. 2010).

There are no miniature wearable sensors that can monitor multiple chemicals in real time. Real-time miniature devices—such as multianalyte single-fiber optical sensors, sol-gel indicator composites, portable acoustic-wave sensors, thin-film resonators, and surface acoustic-wave array detectors—have sufficient sensitivity but poor chemical selectivity in the presence of interfering chemical vapors. Sensors currently provide only one indirect signal, for example, frequency variation due to mass loading or changes in refractive index due to molecular adsorption. Selectivity in detection is often accomplished by immobilizing chemical interfaces or biologic receptors on the sensor surface. Since chemical selectivity is accomplished by using chemically-specific interfaces immobilized on sensor surfaces, their selectivity is only as good as the selectivity of the interfaces and fails when complex mixtures are present (Walt 2005).

Small devices (for example, having roughly the size and appearance of cellular telephones) that could provide some temporal information (such as 5-minute to 1-hour averages) and be downloaded remotely, could be attractive alternatives to personal and microenvironmental monitors. The ability to include multiple agents in a given medium would allow fewer measurements to be taken, minimizing the time for setting up and collecting samples in homes and the number of samplers worn or carried by a single person. A universal platform for measuring compounds of interest, or at least screening samples for those more likely to have high concentrations, would have a major influence on the cost of assessing exposures and allow more rapid identification of "highly exposed" people to help to identify sources, factors influencing exposures, and means of reducing exposures.

The recent advances in microlithographic technologies and microfabrication and nanofabrication enable the development of smart sensors and devices that can also be mass-manufactured in a cost-effective and modular fashion (Cheng et al. 2006). These advances could be exploited and coupled with advances in electronics and computing in developing convenient, sensitive, and cost-effective physical, chemical, and biologic detection devices. Advances in nanoscience and technology offer an unprecedented opportunity for developing very small integrated sensors. Examples of nanosensor platforms include nanowires, nanoelectromechanical and microelectromechanical systems (NEMS and MEMS), optical resonators, nanoparticles, graphenes, and doped quantum dots (Khanna 2012). All those have unprecedented sensitivity (Shelley 2008). One of the advantages of nanosensors is that multiple sensor elements can be fabricated or incorporated on the same chip for monitoring multiple analytes simultaneously. Nanosensors also allow monitoring of multiple signals—for example, frequency variation due to mass loading and variation in the electric, mechanical, thermal, optical, and magnetic properties due to molecular adsorption—and so can provide increased chemical selectivity. These analytic-induced signals can be orthogonal, allowing pattern-recognition algorithms to identify a chemical selectively.

The most important and desirable characteristics of a sensor include high selectivity, high sensitivity, real-time detection, broad dynamic range, ability to regenerate in a short time, miniature size, ability to detect multiple analytes simultaneously, low power consumption, and low cost. For nanosensors that are required for field deployment, wireless transmission of signals to another location is important.

Chemical Selectivity

Although nanosensors have demonstrated extremely high sensitivity, their selectivity poses a challenge. Most of the chemical sensing of small molecules in the vapor phase is carried out indirectly with physical sensors modified with chemical interfaces or biologic receptors. The receptors or chemical interfaces, immobilized on physical sensors (nanosensors included), adsorb particular analyte molecules with higher affinity than other molecules that impinge on the receptor surface. Examples of the receptors or chemical interfaces include self-assembled monolayers, polymer films, and biomolecules.

Research using microsensors has demonstrated that small molecules can be detected with high sensitivity and specificity in the absence of other interfering molecules. The microsensor surface is immobilized with chemically selective interfaces for molecular recognition. A pattern-recognition algorithm using responses from an array can identify the chemical with certainty. However, the confidence level of pattern recognition decreases with binary mixtures and fails for mixtures of three or more constituents (Hsieh and Zellers 2004; Jin and Zellers 2008; Senesac and Thundat 2008). The failure is related to the weak molecular interactions between analyte and receptor. Selectivity between analyte and receptor will remain a formidable challenge provided that there is an absence of highly selective coatings that can provide a unique signal in an array format. Development of chemical interfaces or receptors that can provide unique signals is important for the advancement of personal monitors.

However, the ability of microsensors and nanosensors to produce signals that are independent of adsorption chemistry potentially leads to enhanced chemical selectivity. A chemical sensor that is based on detecting thermal changes due to infrared excitation of adsorbed molecules and uses a bimaterial cantilever (photothermal deflection spectroscopy) that can detect a monolayer of adsorbed molecules has been demonstrated (Krause et al. 2008). Infrared absorption peaks are unique for specific molecules, so multiple analytes can be detected with pattern-recognition algorithms. Nanocantilever sensors with unprecedented sensitivity have been demonstrated (Li et al. 2007). Combining the nanocantilever with photothermal deflection spectroscopy could provide high selectivity and sensitivity. However, miniature tunable light sources, such as quantum cascade lasers, in the mid-infrared region that can be incorporated into nanocantilevers are still under development (Capasso 2010).

Incorporating nanocantilevers with high-performance GC also has the potential for increasing chemical selectivity. For example, the ability of a cantilever to measure adsorption energy due to molecular binding (cantilever bending) combined with measurement of adsorbed mass (resonance frequency) and high-performance GC can enhance chemical selectivity.

The committee envisions that future nanosensors, where each sensor provides an orthogonal signal, will be able to detect multiple analytes with an array-based concept. In addition, the different modes of operation of nanosensors, for example, using variations in mass, stress, optical, electric, and thermal properties, induced by the analytes, could be incorporated without increasing the power demand for improving selectivity. The signals would be analyzed by using pattern-recognition algorithms for chemical identification. The nanosensor platform also could be equipped with wireless telemetry for data transmission.

Many challenges need to be addressed to have field-deployable nanosensors that can detect multiple chemicals with selectivity, sensitivity, and site location in a continuous fashion for weeks. Selectivity may be the most challenging task. Current nanosensors have wide variability in their response because of uncertainties in the fabrication process and variability in the immobilization of functional groups (Patel et al. 2003; Senesac and Thundat 2008). In addition, data quality issues, including maintenance, calibration, and reference checks will be important.

Because nanosensors are point sensors, they require the analyte molecules to adsorb onto the sensor surface to produce a signal. Therefore, collecting the analyte molecules through exposure of the sensor element is important. In general, adsorption by diffusion, which eliminates the need for a pump, will reduce the size and power requirement of the sensors.

In sum, advances in nanotechnology and nanosensors offer sensor and sample collector platforms that could be used for developing miniaturized, low-power personal monitors for quantitative measurement of particles and other pollutants in real time with specificity. Sample collection and separation could be accomplished with electrostatic methods, and detection could be achieved with NEMS and MEMS mass sensors. It will be possible to incorporate nanosensors that can detect the activity level of the wearer (using breathing rate and heart rate). Present technology allows inclusion of GPS for instantaneous location determination.

Sensors for Ecosystem Exposure Assessment

Scientific and technologic advances in exposure assessment in nonhuman species have been driven largely by environmental laws and policies administered by various state and federal agencies, but these advances have typically been enabled by advances in exposure and health assessments for humans.

Three recommendations from the Environmental Protection Agency (EPA) Science Advisory Board Workshop on Ecological Risk Assessment related to exposure assessment were to increase resolution and decrease uncertainties in spatial, temporal, and individual-level exposure assessments (Dale et al. 2008). The challenge for all three recommendations is that the size, scale, and duration of traditional approaches to exposure assessment needed to accomplish the improvements will be cost-prohibitive. However, recent advances in electronic miniaturization and data management will allow the development of environmental sensor networks that can provide long-term, real-time exposure monitoring data on many scales. The development of both fixed and mobile sensor arrays has become more common, and many systems are beginning to be deployed. For example, the National Ecological Observatory Network (NEON) will provide long-term ecosystem monitoring on a continental scale (Keller et al. 2008), and the Global Lakes Ecological Observatory Network (GLEON) aims to use lakes as sentinels of global climate change with existing and newly developed arrays of sensors deployed in lakes worldwide (Kratz et al. 2006; Williamson et al. 2009).

Multiscale systems could be deployed to conduct low-resolution monitoring to detect areas of interest, where higher-resolution or higher-density sensor arrays would be deployed to increase resolution and frequency of measurements (for example, Rundel et al. 2009). Driven largely by the needs of national security (for example, monitoring drinking water or air quality) and those of programs like NEON and GLEON, the development of new sensors is rapid. That is especially true for sensors based on molecular or biochemical reactions (biosensors), which show great promise for use in monitoring the exposure of ecologic resources to stressors. In addition, the European Global Monitoring for Environment and Security program will soon begin to launch missions to deploy high resolution, multi-spectral sensors that will be used to monitor environmental stressors from the local to the global scale (Aschbacher and Milagro-Pérez

2012). As improvements are made in data management and storage infrastructure on the Internet, sensor networks will play a key role in advancing the science of exposure assessment in the environment.

Although promising, issues of validation for ubiquitous and participatory sensing networks will remain a key issue. If exposure assessments are to be conducted using sensing networks and nanosensors, greater assurance about measurement accuracy and precision will be needed. Such efforts invariably will include laboratory and field testing against "gold" standard instruments.

Conclusions

Challenges and research needs are presented by ubiquitous networks, including embedded and participatory networks for humans and ecosystems. The field is in its infancy, and much work remains to be done, including the following:

- Develop sensor technologies that can be scaled up to large mass markets with little additional cost, and develop the software to process the information on the sensor platform (for example, cellular telephones or radio sensors) or to send it wirelessly.
- Validate ubiquitous sensors against gold standard instruments in human populations.
- Assess measurement error in cellular telephones or radio sensors in comparison with high-caliber instruments by using either laboratory or field analyses.
- Develop robust sensors and sensor platforms that function with little maintenance and can tolerate harsh environmental conditions over long periods in remote locations.
- Develop the database infrastructure to store, maintain, and protect information.
- Create the analytic methods needed to understand patterns, detect outliers, and ultimately address the potentially enormous quantities of observations that come in streams or in real time; all of which will require considerable investment in methodologic research.
- Continue to assess the optimal way of obtaining user participation.
- Address concerns about individual privacy protection so that users understand any potential risks to their privacy and identity.

BIOMONITORING FOR ASSESSING INTERNAL EXPOSURES

The Opportunity

Exposure science is poised to move from collection of external exposure information on a small number of stressors, locations, times, and individuals to a more systematic assemblage of internal exposures of individuals in entire populations and multiple elements of the ecosystem to multiple stressors (Wild 2005; Dale et al. 2008). By providing detailed exposure information to complement the rapidly evolving individualized biomedical profile based on gene and gene-expression data (Cohen et al. 2011; Auffray et al. 2012; Chen et al. 2012), global-exposure surveillance has the potential to become a valuable component of routine health care (see illustration in Box 5-6). Advances in ecologic genomic techniques and bioinformatics provide a basis for conducting routine exposure assessments of broad arrays of stressors in wildlife. The use of those techniques in regulatory ecotoxicology and ecologic risk assessment has limitations (Ankley et al. 2008), but Ankley et al. (2008) state that "Ongoing research will, in the long term, serve to obviate limitations related to the global identification of gene products, proteins, and metabolites in test species relevant to ecological risk assessments." Global approaches, the use of genomewide assessments of indicators of exposure to a wide variety of stressors, will continue to complement traditional targeted measures of internal exposure.

Measures of Internal Exposure

For both ecologic and human health risk assessment, internal measures of exposures to stressors are closer to the target site of action for biologic effects than external measures, and this potentially reduces confounding and improves the correlation of exposures with biologic effects. However, use of internal measures of exposure comes with a cost: the variability in the relationship between sources of stressors and effects is greater than that with the use of external measures of exposure. Nevertheless, the committee considers it important to advance measurement of internal exposures as an element of the vision for exposure science.

Analytic methods enable detection of both much lower concentrations of stressors internally and measurement of multiple stressors in single samples. The global measurement of thousands of small organic molecules (Nicholson and Lindon 2008) in biologic samples—metabolomics—applied initially to biomedical fields is now being applied to biomonitoring of chemicals in human and wildlife populations (Ankley et al. 2008; Stahl et al. 2010; Villenueuve and Garcia-Reyero 2011; Soltow et al. in press). Such global approaches have the distinct advantage of not being limited to a chemical or class of chemicals selected in advance and provide broader, agnostic assessment that can identify exposures, potentially improving surveillance and elucidating emerging contaminants. Approaches specific to chemical classes are becoming equally powerful; for example, the simultaneous determination of 50-77 polychlorinated biphenyls (PCBs) (Korrick et al. 2000; Bloom et al. 2009) and 28 polychlorinated dibenzodioxin, polychlorinated dibenzofuran, and dioxin-like PCB congeners (Todaka et al. 2010) is now routine, as is the measurement of the steroidome—69 steroid hormones in human blood (Hill et al. 2010). Proteomics and adductomics expand the types of internal measures of exposure that can be analyzed to include compounds with short half-lives in the blood, including the critical class of reactive electrophiles, for example, oxidants in cigarette smoke (Tang et al. 2010; Jin et al. 2011) and acrylamide, glycidamide, and styrene oxide (Fustinoni et al. 2008; Feng and Lu 2011). Whether the biologic fluid is blood, urine, or saliva, rapidly evolving sensor platforms linked to physiologically based pharmacokinetic (PBPK) models are expected to enable field measurements in humans and in ecosystems and rapid interpretation of concentrations of chemicals in these biofluids in the context of internal exposure (Timchalk et al. 2007). However, inferring the sources and routes of these internal exposures remains a research challenge (Tan et al. 2007).

Biosignatures of Exposure

An alternative to global surveillance of internal exposure to specific stressors is the use of biosignatures that reflect the net biologic effect of internal exposure to stressors that act on a specific biologic pathway. For example, oxidative modifications of DNA or protein (Zhang et al. 2010) can be used to represent the net internal exposure to oxidants and antioxidants, the presence of liver enzymes in blood may reflect internal exposure to liver toxicants (Shi et al. 2010), induction of the cytochrome P4501A gene pathway reflects exposure to planar aromatic hydrocarbons in humans and wildlife (Nebert et al. 2004), and the presence of the egg-yolk protein precursor vitellogenin in juvenile or male fish can be used to assess exposure to estrogenic compounds in the laboratory (for example, Scholz and Mayer 2008) or in the field (for example, Kidd et al. 2007; Sanchez et al. 2011).

BOX 5-6 Potential Application of -omics and Exposure Data in Personalized Medicine

Applications of combined exposure–disease -omics methods in personalized medicine provide a conceptual framework for use of exposure surveillance in health management. For example, if a cost-effective global analysis were used in routine health care, children exposed to second-hand smoke could be readily identified by detection of cotinine. An important transition might occur if such a test were extended to detect a broad array of exposures to contamination in well water or to household pesticides simultaneously by using laboratory-on-a-chip or chemical profiling. Increasing use of -omics methods in personalized medicine means that integration of exposure science with personalized medicine could allow the cost of a top-down approach to exposure surveillance to be borne largely by the health-care system with systematic acquisition of information on exposures in routine health care without considerable additional cost. Such analyses would require considerable expertise of physicians to interpret and to communicate the health significance of this exposure information to patients.

Using biosignatures has several advantages. Biosignatures can overcome the analytic and informational challenges of identifying all stressors of a common biologic pathway and understanding their individual and summed potencies. The identities of the stressors do not need to be known in advance; rather, their presence is inferred from disruption of a specific biologic pathway. The close connection between the exposure measure and the adverse effect or disease process allows better exposure–disease correlations. A major challenge of the approaches is that using complicated, dynamic biologic systems to measure internal exposure will require overcoming substantial variability in response, temporal variability, the deconvolution of biologic processes into influences from external vs internal stressor-related processes, and the relative inability to target reduction of any specific compound or source on the basis of biosignatures.

Biochemical Modifiers of Internal Exposure

Adsorption, distribution, metabolism, and elimination—which contribute to the relationship between external and internal exposures—are themselves the results of biochemical and physical processes. The state of those processes, as they are related to a stressor of concern, has a substantial effect on the internal exposure that receptors experience. Knowledge of the processes is needed to characterize or predict internal exposures in humans and environmental systems. The committee envisions the extension of traditional biomonitoring to include the monitoring of the processes at the individual level and the population level through emerging technologies.

Transcriptomics, proteomics, and to a smaller extent metabolomics offer the ability to measure the status of key biologic processes that affect the pharmacokinetics of chemical stressors across time, species, and populations. And that information can be used qualitatively to identify populations expected to have greater internal exposures to a given external exposure (for example, because of differences in metabolism or higher absorption) or quantitatively by inclusion in PBPK–pharmacodynamic (PBPK-PD) models that are used for exposure assessment and prediction of doses. PBPK models have been widely used in risk assessment to predict the dose to a target tissue from external exposures (Clewell and Andersen 1987; Teeguarden et al. 2005; Clewell and Clewell 2008) and to address the effect of individual and population variability. For example, PBPK–PD models describe the induction of Phase I and Phase II metabolism in the liver and other tissues to account for the effects of increased metabolism from chronic exposures on internal exposure to one or more compounds (Sarangapani et al. 2002; Emond et al. 2006). While the Bayesian, Monte Carlo, and other computational approaches for applying PBPK models to population-level exposure assessments are well developed, the limited availability of population-level data on variability in external exposures and on individual genetic variation, hinders consistent application to populations. The committee's vision for exposure science is anticipated to help motivate the generation of the data needed to support wider application of these computational approaches.

Ecologic Exposure Assessment

Advances in molecular biology and a desire to provide high-throughput assessment of exposures of organisms in the natural environment have led to the development of numerous tools to provide surrogate or direct measures of exposure to contaminants (Table 5-1). The expansion of biochemical and molecular measures has been particularly rapid. However, the use of molecular techniques as biomarkers to assess ecologic exposure to contaminants is limited in that most of these techniques cannot be linked quantitatively to the level of exposure and are not highly selective. Most biomarkers can provide information on whether exposure to general classes or groups of stressors has occurred and, given the current state of the science, cannot be related causally to susceptibility or to risk of developing disease in natural populations (Stahl et al. 2010). Whereas the qualitative information has been useful in directing the need for higher-resolution studies and focused analytic-chemistry measures (for example, Roberts et al. 2005; Smith et al. 2007), it does not in isolation provide exposure-assessment information that can be useful in ecologic risk assessment.

Conclusions

Monitoring of ecologic resources for exposure to stressors can play many roles, including determination of stressor presence, stressor type, and stressor level. With few exceptions (for example, analytic chemistry), however, monitoring techniques do not provide the quantitative information needed for use in ecologic risk assessments.

- There is a need to develop rapid-response, quantitative exposure-assessment tools that can provide information useful for exposure assessment in ecologic risk assessments.
- The use of molecular techniques as biomarkers to assess ecologic exposures to contaminants has been the subject of much attention and development in recent years. However, the uses of these techniques is limited in that most cannot be linked quantitatively to the level of exposure and to the level producing an adverse outcome. Efforts need to be made to make biomarkers more quantitative.
- Linking quantitative analytic chemistry and biochemical adverse-outcome pathways with biomonitoring tools will be essential to move ecologic exposure science into the 21st century.

TABLE 5-1 Available Methods and Their Utility for Ecologic Exposure Assessment[a]

Method	Stressor Specificity: (H)igh, (M)edium, (L)ow	(Quant)itative, (Semiquan)titative, (Qual)itative	(D)irect or (I)ndirect measure of stressor	Level of Develoment, Acceptability (H)igh, (M)edium, (L)ow	Relative Speed of Analysis	Clear Link to Ecologic Risk-Assessment Goals	Example
Analytic chemistry	H	Quant	D	H	Slow	Yes	Genualdi et al. 2011
Biomimetic devices	H	Semiquant	D	M-H	Slow	Yes	Esteve-Turrillas et al. 2007
Whole-organism biomonitoring	H	Quant	D	H	Slow	Yes	Finkelstein et al. 2007
Integrated stressor assessments	M	Qual	I	M/H	Slow	Yes	Cvetkovic and Chow-Fraser 2011
Sensor networks	H	Quant	D	L	Fast	No	Pastorello et al. 2011
Biomarkers	M	Qual	I	M	Fast	No	Roberts et al. 2005
-omics techniques	M	Qual	I	M	Fast	No	Klaper et al. 2010
Remote sensing	L	Qual	I	M	Medium	No	Yuan et al. 2010; Whitehead et al. 2011
Participatory assessment	L	Qual	I	L	Medium	No	Kolok et al. 2011

[a]These methods range from direct analytic-chemistry measurements of specific chemicals to the use of remote-sensing technologies (for example, satellite imagery). Each method has advantages and disadvantages related to stressor specificity, quantitation of exposure, connectivity of a measurement to a specific stressor, level of scientific development and acceptability, speed (and hence cost) of a measurement, and the ability to link a measurement to scientific or regulatory goals of ecologic risk assessment.

MODELS, KNOWLEDGE, AND DECISIONS

Models of the processes, dynamics, and distribution of exposures to chemical, physical, and biologic agents in association with other stressors are an essential element of exposure science. The ability of models to provide a repository for exposure knowledge, to aid in interpreting data and observations, and to provide tools for predicting trends has been and will continue to be a cornerstone of exposure science. Here we consider the role that exposure models need to play in supporting exposure science in the 21st century.

Types of Models in Exposure Science

The types of models used in exposure science vary widely. Activity-based models track the history of individuals and populations through multiple environments using activity information, whereas process-based models track the movement of chemicals and other stressors from a source to a target (receptor).

The Stochastic Human Exposure and Dose Simulation (SHEDS) model is a widely used activity-based model (EPA 2012a). It simulates individual exposure patterns by using the Consolidated Human Activity Database (CHAD), simulating a person's contact with environmental concentrations probabilistically, estimating the person's exposure–time profile for multiple pathways, and applying Monte Carlo sampling to simulate population exposure (Zartarian et al. 2000). SHEDS estimates exposures of specific subpopulations (for example, children) on a national scale. SHEDS is among a class of models that was developed to carry out rapid exposure characterization of numerous stressors, often in the context of policy analyses, when alternative controls were being examined. In such situations, when screening assessments are needed in counterfactual situations, exposures must be modeled.

There are different types of process-based models. For example, source-to-dose models (that is, models that link environmental fate and transport, exposures, and pharmacokinetics) relate exposures, both conceptually and mathematically, at any biologic level to exposures at any other level or to dose (McKone et al. 2007; Georgopoulos et al. 2009). The challenge is to be able to model from dose to source or source to dose, by using newly developed internal and external markers of exposures (see Figure 5-1).

In addition there are also process-based models (or mass-balance models) that have been developed for screening potential exposures, to ecosystems and humans, at local and global scales. The development of these models has been motivated in part by the need for more accurate characterizations of chemicals transported over regional, continental, and global scales—with a focus on the impacts to both humans and ecologic receptors (MacLeod et al. 2010). Such motivations have stemmed from widespread observations of organic pollutant compounds in vegetation, soil, animals, and human tissues. These process-based models use the principles of mass balance and chemical thermodynamics to track the fate of chemical, physical, and biologic stressors, in indoor environments, and on regional and global scales. Box 5-7 examines the use of large-scale process models to assess human and ecologic exposure potential with regards to long-range transport and persistence.

Some process-based models have high spatial resolution, others focus on regional mass-balance methods. Some are deterministic and attempt to capture a small number of representative scenarios, others are stochastic and probabilistic and attempt to capture the uncertainty and variability of model inputs. There is a growing need for structure–activity models that can classify chemicals with regard to exposure and health-effects potential.

The use of computational exposure models within the regulatory decision process at the EPA and other regulatory agencies continues to grow (NRC 2007). According to NRC (2007):

"This growth is in response to greater demands for quantitative assessment of regulatory activities, including analysis of how well environmental regulatory activities fulfill their objectives and at what cost. Models are essential for estimating a variety of relevant characteristics—including pollutant emissions, ambient conditions, and dose—when direct observation would be inaccessible, infeasible, or unethical."

Predictive exposure models have become particularly important in risk, life-cycle, and sustainability assessments, where there is a need for rapid exposure assessment (NRC 2007, 2009). One concern that arises in using exposure models is the reliability of the exposure estimates. This has resulted in a demand for a more formal treatment of uncertainty in exposure models using qualitative methods and quantitative methods such as Monte Carlo analysis (IPCS 2008; NRC 2009). Although widely endorsed, the use of Monte Carlo methods in exposure modeling remains constrained because of insufficient information on many parameters affecting exposure.

The goals of exposure models include explaining observations, guiding data collection, identifying new questions, bounding outcomes within plausible ranges, predicting exposures of individuals for epidemiologic studies, and illuminating key uncertainties and sensitivities. As discussed earlier, models may also describe internal exposures. Models will continue to provide training, educate the public, and link exposure results more closely to risk assessment and risk-management decisions.

Future of Modeling Methods and Technologies

To support the emerging technologies, the committee focused on several modeling approaches that will be needed to extract useful information from the masses of data that will be generated.

> **BOX 5-7** Global-Scale and Regional-Scale Models Used to Assess Human and Ecologic Exposure Potential in Terms of Long-Range Transport Potential and Persistence
>
> Global scale mass-balance and exposure models have supported development of the concepts of persistence and long-range transport—key hazard indicators used in chemical assessment (Fenner et al. 2005). The multimedia-mass-balance models initially focused on a "unit" or "evaluative" world approach in which mass balance is applied to an archetypal world made up of soil, water, sediments, and biota (MacLeod et al. 2010). Those types of evaluative models were instrumental in elucidating the mechanisms of bioconcentration and biomagnification (McKone and MacLeod 2003). Global-scale multimedia models illustrate the potential for some persistent pollutants to migrate to and accumulate in polar regions (Wania and Mackay 1993; Scheringer et al. 2000; Wania and Su 2004). Multimedia mass-balance models provide a basis for quantifying cumulative, multipathway exposure to pollutants that originate in contaminated air, water, and soil (McKone and MacLeod 2003). For example, they have been used to characterize the relative importance of local and distant sources of contaminants for human and ecosystem exposures (MacLeod et al. 2002), to identify environmental and chemical processes that control concentrations of pollutants (Scheringer et al. 2004), and to motivate empirical and theoretical studies to improve knowledge of the physicochemical properties and degradability of commercially important chemicals and byproducts of industry and energy production (Jaworska et al. 2003; Schenker et al. 2005).
>
> More detailed global mass-balance models have been introduced in which detailed spatial differentiation is obtained by linking unit world models with flows of air and water estimated on the basis of long-term averages (MacLeod et al. 2010). Wania and Mackay (1999) argued that the unit world structure and generally low spatial and temporal resolution of multimedia models are appropriate for describing the behavior of persistent contaminants in the environment. However, limitations of that approach have been identified by developers of contaminant fate and transport models. Those models describe the temporal and spatial variability of the circulation of the atmosphere and oceans with high resolution (Lammel 2004). Such detailed chemistry-transport models are increasingly being developed or adapted to describe concentrations and transport pathways of trace environmental contaminants. They have been used to address aspects of chemical pollution that cannot be readily addressed by models that are based on the unit world approach, such as how persistence and long-range transport of substances depend on the specific time and location of release, and how episodic transport events can deliver pollutants to remote ecosystems (MacLeod et al. 2010).

Model Performance Evaluation

Key to the future of exposure models is how they incorporate the ever-increasing amounts of observations of natural and human processes and environmental effects. Vast new measurement programs in fields as diverse as genomics and earth-observation systems, ranging from the nanoscale to global dimensions, present important opportunities and challenges for modeling. Although observations alone can influence policy, it is their analysis with models that will allow the full realization of their importance (NRC 2007).

The interdependence of models and measurements is complex and iterative (NRC 2007; EPA 2009). Historically, the cycle of models and observations tended to begin with observations used to build a model, then a second set of observations to calibrate the model, and a third set to "validate" or evaluate the model performance. But in place of validation or even calibration with observations, an alternative approach is to use models and observations (such as biomarker data and environmental samples) as independent tools to evaluate hypotheses about source–receptor relationships (see McKone et al. 2007).

Spatiotemporal models play a central role in exposure assessment for epidemiologic studies and risk analysis by combining information on sources of exposure with measurements to arrive at predictions of exposures. In some cases, cost and feasibility may limit direct measurements to a modest number of locations for specified periods; in other cases, remote-sensing technologies may yield an enormous quantity of data that need to be summarized.

An important element of any exposure assessment is quantifying the degree of uncertainty about the exposure estimates. It is important to distinguish two fundamentally different sources of uncertainty: statistical variability in the finite set of measurements available on which to build the model and their inherent measurement errors, and misspecification of the form of the statistical model used—both in prediction of exposure and in the sampling and measurement error distributions.

The first of those relies on statistical inference. Well-established techniques are available to estimate the standard errors of parameter estimates and model predictions. The measurement-error literature distinguishes two principal classes of measurement-error models on the basis of the relationship between a "true exposure" and a "measured exposure". In both, the health outcome is assumed to depend only on true exposure, not its measurement; the relationship between the outcome and true exposure, commonly known as the health-effects model, and the population distribution of true exposure (Richardson and Ciampi 2003) (possibly in relation to other predictors, such as traffic density, land use, and other exposure sources) is known as the exposure model. The models differ in terms of the form of the "measurement-error model". In the "classical error model", the measured value is assumed to be distributed randomly around the true value with some deviations caused by instrument or model error. In the "Berkson error model", individual exposures are assumed to be distributed around some applied exposure of a group (for example, ambient pollution). The "classical error model" is an appropriate model for situations in which personal measurements that are subject to some error are used. It tends to induce an attenuation bias toward the null in the "naïve" regression of outcome on measured exposure. The "Berkson error model" is appropriate for situations in which individuals within a group differ because of unmeasured factors (for example, time–activity or household characteristics); it might not produce any bias but will tend to inflate the variance of the estimated regression coefficient. In many instances, a combination of components with characteristics of classical and Berkson error may affect the estimated coefficients.

The second type of uncertainty, model misspecification, is more difficult (Leamer 1978) because the true form of the relationships being modeled can never be known and the most one can hope to do is to investigate the robustness of the predictions in a reasonable group of plausible alternative models. This is generally known as sensitivity analysis. But more formal alternatives, such as model averaging (Hoeting et al. 1999; Hjort and Claeskens 2003), are available that essentially provide an average of a wide variety of models, each weighted in some fashion according to goodness-of-fit, and an overall estimate of uncertainty that combines the statistical variability within each model and the differences between models to provide a more "honest" assessment of overall uncertainty than simply quoting the standard error of some "best" model.

Models that rely heavily on expert judgment, such as dose-reconstruction models used in environmental radiation-exposure studies—for example, the Hanford Environmental Dose Reconstruction Project (Shipler et al. 1996; Kopecky et al. 2004) and the Nevada Test Site down-winders studies (Simon et al. 1995)—have tended to use Monte Carlo methods for uncertainty assessment (see discussion in Box 3-2). In this case, one typically samples random values for each of the unknown parameters from their uncertainty distributions, computes exposure predictions by using these values, and repeats this process many times to generate a distribution of possible exposure estimates.

Uncertainty estimates in exposure have little utility by themselves unless they are exploited in the analysis of exposure–response relationships. Many statistical methods have been developed for dealing with this problem (Thomas et al. 1993; Carroll et al. 2006), but their use in epidemiologic and ecologic analyses has been fairly narrow. For example, in radiation epidemiology, use of the Monte Carlo uncertainty estimates in analyzing health effects has become somewhat standard, as in the analysis of the Hanford Thyroid Disease Study (Stram and Kopecky 2003). In air pollution, Bayesian methods that entail fitting the exposure and health-outcome models jointly have been discussed (Molitor et al. 2007).

Information-Management Tools

In addition to models as tools to interpret, predict, and evaluate source, concentration, and receptor relationships, there are informatics models—the emerging tools for managing and exploring massive amounts of information from diverse sources and in widely different formats.

Substantial investment and progress have occurred in recent years to collect and improve access to genomic, toxicology, and health data. For example,

EPA's ACTOR database (EPA 2012b) "is an on-line searchable warehouse of all publicly available chemical-toxicity data. It aggregates data from over 500 public sources on over 500,000 environmental chemicals; the data are searchable by chemical name, other identifiers, and chemical structure" (Judson et al. 2012). The Comparative Toxicogenomics Database (CTD 2012) entails literature curation and integration of data that describe chemical interactions with genes and proteins and chemical–disease and gene–disease relationships (Mattingly 2009; Davis et. al. 2011). Those databases and efforts to link and integrate the information that they contain have revealed the landscape of chemical-toxicity data and helped to inform research needs. The data sources have historically lacked the extensive and reliable exposure data that are required to examine the environmental contributions to diseases and to assess health risks. EPA initiated the ExpoCast™ program to fill that knowledge gap (Cohen Hubal et al. 2010). Although the program was developed to advance characterization of exposure to meet challenges posed by new toxicity-testing paradigms, in the long run it will foster novel exposure-science research to link exposures to real-world health outcomes.

ExpoCastDB, developed as a component of ExpoCast™ (Gangwal 2011), has begun the effort to consolidate observational human exposure data, improve access, and provide links to health-related data. It is designed to house measurements from human exposure studies and encourage standardized reporting of observational exposure information (EPA 2012c). This database will facilitate linkages with many other data sources, including sources of toxicity and environmental-fate data, and with manufacturers' production and use data.

In the fields of biology and medicine, the need to manage exponentially growing information on systems and interactions has given rise to the discipline of bioinformatics. Bioinformatics brings together model algorithms, databases and information systems, Web technologies, artificial intelligence, and soft computing to generate new knowledge of biology and medicine. One example is Genome-Wide Associated Studies (GWAS), which explores how noncommunicable diseases arise from a complex combination of genetic processes and the environment (Schwartz and Collins 2007). The application of informatics to exposure science is at a very early stage and offers a substantial opportunity to gain new knowledge of stressor–receptor–effect patterns—in connection with humans and the environment.

The fields of systems biology and exposure biology have been a staging area for some of this effort. One example is that of work by Patel et al. (2010) who conducted a pilot Environment-Wide Association Study (EWAS) in which exposure-biomarker and disease-status data were systematically interpreted in a manner analogous to that in GWAS (see Box 3-3).

Fields that make use of informatics technologies often develop an ontology to guide the systematic mining of databases and scientific literature that contain millions of observations and findings. An ontology is an explicit formal specification of the terms used in a knowledge domain (for example, biology, chemistry, or toxicology) and the relations among them. Because "an ontology defines a common vocabulary for researchers who need to share information in a domain" (Noy and McGuinness 2001), it can be adapted to machine-interpretable definitions of basic concepts to facilitate the use of computers for mining of large amounts of written material. Mattingly et al. (2012) recently designed, developed, and demonstrated an exposure ontology (ExO) to facilitate centralization and integration of exposure data. The ExO is being used to bridge the gap between exposure science and other environmental health disciplines, including toxicology, epidemiology, disease surveillance, and epigenetics (OBO Foundry 2012). The committee sees these types of efforts as important for defining and expanding exposure science in the 21st century.

Conclusions

The data and analytic challenges posed by the emerging exposure-assessment technologies are growing at an exponential rate in scale and complexity.

- Statistical and computational research must continue to expand to keep pace with these developments. For example, satellite imaging and personal-monitoring techniques are generating enormous quantities of data on spatiotemporal exposure and on people's movements and activities. Meanwhile, biologic assays are capable of monitoring millions of genetic variants, metabolites, expressions, and epigenetic changes in thousands of subjects at an affordable cost.
- Without comparable investments in the development of new statistical analytic techniques to address correlated data on many more variables than subjects and without advances in computation, such as parallel-processing techniques, the analysis of the mountains of data threatens to become the new limiting factor in further progress.
- Exposure models will continue to support diverse efforts, such as risk analysis, impact assessments, life-cycle and sustainability assessments, epidemiology, and energy analysis. As is the case with all models, exposure models must balance the need for transparency with the need for fidelity and credibility. That requires concurrent development of model performance-evaluation efforts in any model-development program.

REFERENCES

AethLabs. 2012. microAeth Model AE51. AethLabs, San Francisco, CA [online]. Available: http://www.aethlabs.com/ [accessed Jan. 11, 2012].

Allen-Piccolo, G., J.V. Rogers, R. Edwards, M.C. Ciark, T.T. Allen, I. Ruiz-Mercado, K.N. Shields, E. Canuz, and K.R. Smith. 2009. An ultrasound personal locator for time-activity assessment. Int. J. Occup. Environ. Health 15(2):122-132.

Almanza, E., M. Jerrett, G. Dunton, E. Seto, and M.A. Pentz. 2012. A study of community design, greenness and physical activity in children using satellite, GPS and accelerometer data. Health Place 18(1):46-54.

Ankley, G.T., A.L. Miracle, E.J. Perkins, and G.P. Daston. 2008. Genomics in Regulatory Ecotoxicology: Applications and Challenges. Pensacola, FL: SETAC Press.

Aschbacher, J., and M.P. Milagro-Pérez. 2012. The European Earth monitoring (GMES) programme: Status and perspectives. Remote Sens. Environ. 120:3-8.

Auffray, C., T. Caulfield, M. J. Khoury, J.R. Lupski, M. Schwab, and T. Veenstra. 2012. Looking back at genomic medicine in 2011. Genome Med. 4(1):9.

AVIRIS (Airborne Visible/Infared Imaging Spectrometer). 2012. AVIRIS. Jet Propulsion Laboratory. California Institute of Technology[online]. Available: http://aviris.jpl.nasa.gov/ [accessed May 17, 2012].

Bagheri, S., and T. Yu. 2008. Hyperspectral sensing for assessing nearshore water quality conditions of Hudson/Raritan estuary. J. Environ. Inform. 11(2):123-130.

Beck, L.R., B.M. Lobitz, and B.L. Wood. 2000. Remote sensing and human health: New sensors and new opportunities. Emerg. Infect. Dis. 6(3):217-227.

Bloom, M.S., J.E. Vena, J.R. Olson, and P.J. Kostyniak. 2009. Assessment of polychlorinated biphenyl congeners, thyroid stimulating hormone, and free thyroxine among New York state anglers. Int. J. Hyg. Environ. Health 212(6):599-611.

Bradley, E.S., D.A. Roberts, P.E. Dennison, R.O. Green, M. Eastwood, S.R. Lundeen, I.B. McCubbin, and I. Leifer. 2011. Google Earth and Google Fusion tables in support of time-critical collaboration: Mapping the deepwater horizon oil spill with the AVIRIS airborne spectrometer. Earth Sci. Inform. 4(4):169-179.

Briggs, D. 2005. The role of GIS: Coping with space (and time) in air pollution exposure assessment. J. Toxicol. Environ. Health A 68(13-14):1243-1261.

Bulgarelli, B., and S. Djavidnia. 2012. On MODIS retrieval of oil spill spectral properties in the marine environment. IEEE Geosci. Remote Sens. Lett. 9(3):398-402.

Burke, J., D. Estrin, M. Hansen, A. Parker, N. Ramanathan, S. Reddy, and M.B. Srivastava. 2006. Participatory sensing. Proceedings of ACM SenSys, the 4th ACM Conference on Embedded Networked Sensor Systems, October 31, 2006, Boulder, CO [online]. Available: http://escholarship.org/uc/item/19h777qd#page-1 [accessed Jan. 17, 2012].

Calabrese, F., M. Colonna, P. Lovisolo, D. Parata, and C. Ratti. 2011. Real-time urban monitoring using cell phones: A case study in Rome. IEEE T. Intell. Transp. Syst 12(1):141-151.

Capasso, F. 2010. High-performance midinfrared quantum cascade lasers. Opt. Eng. 49(11):111102.

Carroll, R.J., D. Ruppert, L.A. Stefanski, and C.M. Crainiceanu. 2006. Measurement Error in Nonlinear Models: A Modern Perspective, 2nd Ed. Boca Raton, FL: Chapman and Hall/CRC Press.

Chang, L.T., J. Sarnat, J.M. Wolfson, L. Rojas-Bracho, H.H. Suh, and P. Koutrakis. 1999. Development of a personal multi-pollutant exposure sampler for particulate matter and criteria gases. Pollut. Atmos. 40:31-39.

Chen, C. 2011. A Wireless Hybrid Chemical Sensor for Detection of Environmental Volatile Organic Compounds. M.S. Thesis, Arizona State University.

Chen, R., G.I. Mias, J. Li-Pook-Than, L. Jiang, H.Y. Lam, R. Chen, E. Miriami, K.J. Karczewski, M. Hariharan, F.E. Dewey, Y. Cheng, M.J. Clark, H. Im, L. Habegger, S. Balasubramanian, M. O'Huallachain, J/T. Dudley, S. Hillenmeyer, R. Haraksingh, D. Sharon, G. Euskirchen, P. Lacroute, K. Bettinger, A.P. Boyle, M. Kasowski, F. Grubert, S. Seki, M. Garcia, M. Whirl-Carrillo, M. Gallardo, M.A. Blasco, P.L. Greenberg, P. Snyder, T.E. Klein, R.B. Altman, A.J. Butte, E.A. Ashley, M. Gerstein, K.C. Nadeau, H. Tang, and M. Snyder. 2012. Personal omics

profiling reveals dynamic molecular and medical phenotypes. Cell. 148(6):1293-1307.
Cheng, M.M.C., G. Cuda, Y.L. Bunimovich, M. Gaspari, G.R. Hearth, H.D. Hill, C.A. Mirkin, A.J. Nijdam, R. Terracciano, T. Thundat, and M. Ferrari. 2006. Nanotechnologies for biomolecular detection and medical diagnostics. Curr. Opin. Chem. Biol. 10(1):11-19.
Chowdhury, Z., R.D. Edwards, M. Johnson, K.N. Shields, T. Allen, E. Canuz, and K.R. Smith. 2007. An inexpensive light-scattering particle monitor: Field evaluation. J. Environ. Monitor. 9(10):1099-1106.
Chudnovsky, A., E. Ben-Dor, A.B. Kostinski, and I. Koren. 2009. Mineral content analysis of atmospheric dust using hyperspectral information from space. Geophys. Res. Lett. 36: L15811, doi:10.1029/2009GL037922.
Chudnovsky, A., A. Kostinski, L. Herrmann, I. Koren, G. Nutesku, and E. Ben-Dor. 2011. Hyperspectral spaceborne imaging of dust-laden flows: Anatomy of Saharan storm from the Bodele Depression. Remote Sens. Environ. 115(4):1013-1024.
Clark, R.N., R.O. Green, G.A. Swayze, G. Meeker, S. Sutley, T.M. Hoefen, K.E. Livo, G. Plumlee, B. Pavri, C. Sarture, S. Wilson, P. Hageman, P. Lamothe, J.S. Vance, J. Boardman, I. Brownfield, C. Gent, L.S. Morath, J. Taggart, P.M. Theodorakos, and M. Adams. 2001. Environmental Studies of the World Trade Center Area After the September 11, 2001 Attack. US Geological Survey Open File Report 01-0429-2001. U.S. Geological Survey [online]. Available: http://pubs.usgs.gov/of/2001/ofr-01-0429/ [accessed Jan. 17, 2012].
Clewell, H.J, III, and M.E. Andersen. 1987. Dose, species and route extrapolation using physiologically based pharmacokinetic models. Pp. 159-182 in Drinking Water and Health, Vol. 8. Pharmacokinetics in Risk Assessment. Washington, DC: National Academy Press.
Clewell, R.A., and H.J. Clewell, III. 2008. Development and specification of physiologically based pharmacokinetic models for use in risk assessment. Regul. Toxicol. Pharmacol. 50(1):129-143.
Cohen, A.L., R. Soldi, H. Zhang, A.M Gustafson, R. Wilcox, B.W. Welm, J.T. Chang, E. Johnson, A. Spira, S.S. Jeffrey, and A.H. Bild. 2011. A pharmacogenomic method for individualized prediction of drug sensitivity. Mol. Syst. Biol. 7:513.
Cohen-Hubal, E.A., A.M. Richard, L. Aylward, S.W. Edwards, J. Gallagher, M. Goldsmith, S. Isukapalli, R. Tornero-Velez, E.J. Weber, and D.R. Kavlock. 2010. Advancing exposure characterization for chemical evaluation and risk assessment. J. Toxicol. Environ. Health B Crit. Rev. 13(2-4):299-313.
Crall, A.W., G.J. Newman, C.S. Jarnevich, T.J. Stohlgren, D.M. Waller, and J. Graham. 2010. Improving and integrating data on invasive species collected by citizen scientists. Biol. Invasions 12(10):3419-3428.
CTD (Comparative Toxicogenomics Database). 2012. CTD [online]. Available: http://ctdbase.org/ [accessed Jan. 11, 2012].
Cvetkovic, M., and P. Chow-Fraser. 2011. Use of ecological indicators to assess the quality of Great Lakes coastal wetlands. Ecol. Indic. 11(6):1609-1622.
Cycling Metro Vancouver. 2007. Cycling metro Vancouver [online]. Available: http://www.cyclevancouver.ubc.ca/cv.aspx [accessed Oct. 28, 2011].
Dale, V., G.R. Biddinger, M.C. Newman, J.T. Oris, G.W. Suter, T. Thompson, T.M. Armitage, J.L. Meyer, R.M. Allen-King, G.A. Burton, P.M. Chapman, L.L. Conquest, I.J. Fernandez, W.G. Landis, L.L. Master, W.J. Mitsch, T.C. Mueller, C.F. Rabeni, A.D. Rodewald, J.G. Sanders, and I.L. van Heerden. 2008. Enhancing the ecological risk assessment process. Integr. Environ. Assess. Manag. 4(3):306-313.

Davis, A.P., B.L. King, S. Mockus, C.G. Murphy, C. Saraceni-Richards, M. Rosenstein, T. Wiegers, and C.J. Mattingly. 2011. The Comparative Toxicogenomics Database: Update 2011. Nucleic Acids Res. 39(suppl.1):D1067-D1072.

de Nazelle, A, E. Seto, D. Donaire, M. Mendez, J. Matamala, M. Portella, D. Rodriguez, M. Nieuwenhuijsen, and M. Jerrett. 2011. Improving Estimates of Travel Activity and Air Pollution Exposure through Ubiquitous Sensing Technologies. Abstract S-0035 in Abstracts of the 23rd Annual Conference of the International Society of Environmental Epidemiology (ISEE), September 13-16, 2011, Barcelona, Spain [online]. Available: http://ehp03.niehs.nih.gov/article/fetchArticle.action?articleURI =info%3Adoi%2F10.1289%2Fehp.isee2011 [accessed Sept. 4, 2012].

Demokritou, P., I.G. Kavouras, S.T. Ferguson, and P. Koutrakis. 2001. Development and laboratory performance evaluation of a personal multipollutant sampler for simultaneous measurements of particulate and gaseous pollutants. Aerosol Sci. Technol. 35(3):741-752.

Doran, M., M. Babin, O. Hembise, A. Mangin, and P. Garnesson. 2011. Ocean transparency from space: Validation of algorithms estimating Secchi depth using MERIS, MODIS, and SeaWiFS data. Remote Sens. Environ. 115(12):2986-3001.

Dunton, G.F., Y. Liao, S. Intille, J. Wolch, and M. Pentz. 2011. Social and physical contextual influences on children's leisure-time physical activity: An ecological momentary assessment study. J. Phys. Act. Health 8(suppl. 1):S103-S108.

Eakin, C.M., C.J. Nim, R.E. Brainard, C. Aubrecht, C. Elvidge, D.K. Gledhill, F. Muller-Karger, P.J. Mumby, W J. Skirving, A.E. Strong, M. Wang, S. Weeks, F. Wentz, and D. Ziskin. 2010. Monitoring coral reefs from space. Oceanography 23(4):118-133.

Edwards, R., K.R. Smith, B. Kirby, T. Allen, C.D. Litton, and S. Hering. 2006. An inexpensive dual-chamber particle monitor: Laboratory characterization. Air Waste Manage. Assoc. 56(6):789-799.

Elgethun, K., M.G. Yost, C.T. Fitzpatrick, T.L. Nyerges, and R.A. Fenske. 2007. Comparison of global positioning system (GPS) tracking and parent-report diaries to characterize children's time-location patterns. J. Expo. Sci. Environ. Epidemiol. 17(2):196-206.

Emilli, E., C. Popp, M. Petitta, M. Riffler, S. Wunderle, and M. Zebisch. 2010. PM_{10} remote sensing from geostationary SEVIRI and polar-orbiting MODIS sensors over the complex terrain of the European Alpine region. Remote Sens. Environ. 114(11):2485-2499.

Emond, C., L.S. Birnbaum, and M.J. Devito. 2006. Use of a physiologically based pharmacokinetic model for rats to study the influence of body fat mass and induction of CYP1A2 on the pharmacokinetics of TCDD. Environ. Health Perspect. 114(9):1394-1400.

Engel-Cox, J.A., R.M. Hoff, R. Rogers, F. Dimmick, A.C. Rush, J.J. Szykman, J. Al-Saadi, D.A. Chu, and E.R. Zell. 2006. Integrating lidar and satellite optical depth with ambient monitoring for 3-dimensional particulate characterization. Atmos. Environ. 40(40):8056-8067.

EnMAP (Environmental Mapping and Analysis Program). 2011. EnMAP Hyperspectral Imager [online]. Available: http://www.enmap.org/ [accessed Oct. 28, 2011].

EPA (U.S.Environmental protection Agency). 2009. Guidance on the Development, Evaluation, and Application of Environmental Models. EPA/100/K-09/003. Office of Science Advisor, Council for Regulatory Environmental Modeling, U.S. Environmental Protection Agency [online]. Available: http://www.epa.gov/crem/library/ cred_guidance_0309.pdf [accessed Jan. 25, 2012].

EPA (U.S. Environmental Protection Agency). 2012a. SHEDS-Multimedia: Stochastic Human Exposure and Dose Model for Multimedia, Multipathway Chemicals. Human Exposure and Atmospheric Sciences, National Exposure Research Laboratory, U.S. Environmental Protection Agency [online]. Available: http://www.epa.gov/heasd/products/sheds_multimedia/sheds_mm.html [accessed May 1, 2012].

EPA (U.S. Environmental Protection Agency). 2012b. ACTor. National Center for Computational Toxicology, U.S. Environmental Protection Agency [online]. Available: http://actor.epa.gov/actor/faces/ACToRHome.jsp [accessed Jan. 12, 2012].

EPA (U.S. Environmental Protection Agency). 2012c. ExpoCastDB [online]. Available: http://actor.epa.gov/actor/faces/ExpoCastDB/Home.jsp [accessed June 1, 2012].

Esteve-Turrillas, F.A., A. Pastor, V. Yusa, and M. de la Guardia. 2007. Using semipermeable membrane devices as passive samplers. Trends Anal. Chem. 26(7):703-712.

Feng, C.H., and C.Y. Lu. 2011. Modification of major plasma proteins by acrylamide and glycidamide: Preliminary screening by nano liquid chromatography with tandem mass spectrometry. Anal. Chim. Acta 684(1-2):80-86.

Fenner, K., M. Scheringer, M. Macleod, M. Matthies, T. McKone, M. Stroebe, A. Beyer, M. Bonnell, A.C. Le Gall, J. Klasmeier, D. Mackay, D. Van De Meent, D. Pennington, B. Scharenberg, N. Suzuki, and F. Wania. 2005. Comparing estimates of persistence and long-range transport potential among multimedia models. Environ. Sci Technol. 39(7):1932-1942.

Ferrier, G. 1999. Application of imaging spectrometer in identifying environmental pollution caused by mining at Rodaquilar, Spain. Remote Sens. Environ. 68(2):25-137.

FGDC (Federal Geographic Data Committee). 2011. Standards [online]. Available: http://www.fgdc.gov/standards [accessed May 11, 2012].

Finkelstein, M.E., K.A. Grasman, D.A. Croll, B.R. Tershy, B.S. Keitt, W.M. Jarman, and D.R. Smith. 2007. Contaminant-associated alteration of immune function in black-footed albatross (*Phoebastria nigripes*), a North Pacific predator. Environ. Toxicol. Chem. 26(9):1896-1903.

Focardi, S., I. Corsi, S. Mazzuoli, L. Vignoli, and S.A. Loiselle. 2006. Integrating remote sensing approach with pollution monitoring tools for aquatic ecosystem risk assessment and management: A case study of Lake Victoria (Uganda). Environ. Monit. Assess. 122(1-3):275-287.

Fustinoni, S., L. Campo, P. Manini, M. Buratti, S. Waidyanatha, G. De Palma, A. Mutti, V. Foa, A. Colombi, and S. M. Rappaport. 2008. An integrated approach to biomonitoring exposure to styrene and styrene-(7,8)-oxide using a repeated measurements sampling design. Biomarkers 13(6):560-578.

Gallagher, L.G., T.F. Webster, A. Aschengrau, and V.M. Vieira. 2010. Using residential history and groundwater modeling to examine drinking water exposure and breast cancer. Environ. Health Perspect. 118(6):749-755.

Gangwal, S. 2011. ExpoCastDB: A Publicly Accessible Database for Observational Exposure Data. Presented at Computational Toxicology Community of Practice, September 22, 2011 [online]. Available: http://www.epa.gov/ncct/download_files/chemical_prioritization/ExpoCastDB_CommPractice_09-22-2011-Share.pdf [accessed Dec. 7, 2011].

Genualdi, S.A., K.J. Hageman, L.K. Ackerman, S. Usenko, and S.L.M. Simonich. 2011. Sources and fate of chiral organochlorine pesticides in western U.S. National Park ecosystems. Environ. Toxicol. Chem. 30(7):1533-1538.

Georgopoulos, P.G., A.F. Sasso, S.S. Isukapalli, P.J. Lioy, D.A. Vallero, M. Okino, and L. Reiter. 2009. Reconstructing population exposures to environmental chemicals from biomarkers: Challenges and opportunities. J. Expo. Sci. Environ. Epidemiol. 19(2):149-171.

Goetz, S.J., N. Gardiner, and J.H. Viers. 2008. Monitoring freshwater, estuarine and nearshore benthic ecosystems with multi-sensor remote sensing: An introduction to the special issue. Remote Sens. Environ. 112(11):3993-3995.

Gomez-Casero, M.T., I.L. Castillejo-Gonzalez, A. Garcia-Ferrer, J.M. Pena-Barragan, M. Jurado-Exposito, L. Garcia-Torres, and F. Lopez-Granados. 2010. Spectral discrimination of wild oat and canary grass in wheat fields for less herbicide application. Agron. Sustain. Dev. 30(3):689-699.

Goodchild, M.F. 2007. The Morris Hansen Lecture 2006: Statistical perspectives on spatial social science. J. Off. Stat. 23(3):269-283.

GPS (Global Positioning System). 2011. The Global Positioning System [online]. Available: http://www.gps.gov/systems/gps/ [accessed Oct. 28, 2011].

Gunier, R.B., M.H. Ward, M. Airola, E.M. Bell, J. Colt, M. Nishioka, P.A. Buffler, P. Reynolds, R.P. Rull, A Hertz, C. Metayer, and J.R. Nuckols. 2011. Determinants of agricultural pesticide concentrations in carpet dust. Environ. Health Perspect. 119(7):970-976.

Hagerstrand, T. 1970. What about people in regional science. Pap. Regul. Sci. 24:1-21 (as cited in Briggs 2005).

Hill, M., A. Parizek, R. Kancheva, M. Duskova, M. Velikova, L Kriz, M. Klimkova, A. Paskova, Z. Zizka, P. Matucha, M. Meloun, and L. Starka. 2010. Steroid metabolome in plasma from the umbilical artery, umbilical vein, maternal cubital vein and in amniotic fluid in normal and preterm labor. J. Steroid Biochem. Mol. Biol. 121(3-5):594-610.

Hjort, N.L., and G. Claeskens. 2003. Frequentist model average estimators. J. Am. Stat. Assoc. 98(464):879-899.

Hoeting, J.A., D. Madigan, A.E. Raftery, and C.T. Volinsky. 1999. Bayesian model averaging: A tutorial. Statist. Sci. 14(4):382-417.

Hoff, R.M., and S.A. Christopher. 2009. Remote sensing of particulate pollution from space: Have we reached the promised land? J. Air Waste Manage. Assoc. 59(6):645-675.

Houston, D., P. Ong, J. Wu, and A. Winer. 2006. Proximity of licensed child care facilities to near-roadway vehicle pollution. Am. J. Public Health 96(9):1611-1617.

Hsieh, M.D., and E.T. Zellers. 2004. Limits of recognition for simple vapor mixtures determined with a microsensor array. Anal. Chem. 76(7):1885-1895.

Iglesias, R.A., F. Tsow, R. Wang, E.S. Forzani, and N. Tao. 2009. Hybrid separation and detection device for analysis of benzene, toluene, ethylbenzene, and xylenes in complex samples. Anal. Chem. 81(21):8930-8935.

Intille, S.S. 2007. Technological innovations enabling automatic, context-sensitive ecological momentary assessment. Pp. 308-337 in The Science of Real-Time Data Capture: Self-Report in Health Research, A.A. Stone, S. Shiffman, A. Atienza, and L. Nebeling, eds. Oxford: Oxford University Press [online]. Available: http://www.ccs.neu.edu/home/intille/teaching/AMB/papers/Stone_Chapter16.pdf [accessed Jan. 20, 2012].

IPCS (International Programme on Chemical Safety). 2008. Uncertainty and Data Quality in Exposure Assessment, Part 1. Guidance Document on Characterizing and Communicating Uncertainty of Exposure Assessment. IPCS project on the Harmonization of Approaches to the Assessment of Risk from Exposure to Chemicals.

Geneva: World Health Organization [online]. Available: http://www.who.int/ipcs/publications/methods/harmonization/exposure_assessment.pdf [accessed May 14, 2012].

Jaworska, J.S., R.S. Boethling, and P.H. Howard. 2003. Recent developments in broadly applicable structure-biodegradability relationships. Environ. Toxicol. Chem. 22(8):1710-1723.

Jerrett, M., A. Arain, P. Kanaroglou, B. Beckerman, D. Potoglou, T. Sahsuvaroglu, J. Morrison, and C. Giovis. 2005. A review and evaluation of intraurban air pollution exposure models. J. Expo. Anal. Environ. Epidemiol. 15(2):185-204.

Jerrett, M., S. Gale, and C. Kontgis. 2009. An environmental health geography of risk. Pp. 418-445 in A Companion to Health and Medical Geography, T. Brown, S. McLafferty, and G. Moon, eds. Oxford, UK: Wiley-Blackwell.

Jin, C., and E.T. Zellers. 2008. Limits of recognition for binary and ternary vapor mixtures determined with multi-transducer arrays. Anal. Chem. 80(19):7283-7293.

Jin, H., B.J. Webb-Robertson, E.S. Peterson, R. Tan, D.J. Bigelow, M.B. Scholand, J.R. Hoidal, J.G. Pounds, and R.C. Zangar. 2011. Smoking, COPD and 3-nitrotyrosine levels of plasma proteins. Environ. Health Perspect. 119(9):1314-1320.

Jones, A.P., E.G. Coombes, S.J. Griffin, and E.M. van Sluijs. 2009. Environmental supportiveness for physical activity in English school children: A study using Global Positioning Systems. Int. J. Behav. Nutr. Phys. Act. 6(1):42.

Judson, R.S., M.T. Martin, P.P. Egeghy, S. Gangwal, D.M. Reif, P. Kothiya, M.A. Wolf, T. Cathey, T.R. Transue, D. Smith, J. Vail, A. Frame, S. Mosher, E.A. Cohen-Hubal, and A.M. Richard. 2012. Aggregating data for computational toxicology applications: The U.S. Environmental Protection Agency (EPA) Aggregated Computational Toxicology Resource (ACToR) system. Int. J. Mol. Sci. 13(2):1805-1831.

Keller, M., D.S. Schimel, W.W. Hargrove, and F.M. Hoffman. 2008. A continental strategy for the National Ecological Observatory Network. Front. Ecol. Environ. 6(5):282-284.

Khanna, V.K. 2012. Nanosensors: Physical, Chemical, and Biological. Boca Raton, FL: CRC Press.

Kidd, K.A., P.J. Blanchfield, K.H. Mills, V.P. Palace, R.E. Evans, J.M. Lazorchak, and R.W. Flick. 2007. Collapse of a fish population after exposure to a synthetic estrogen. Proc. Natl. Acad. Sci. U.S.A. 104(21):8897-8901.

Kim, S.K., H. Chang, and E.T. Zellers. 2011. Microfabricated gas chromatograph for the selective determination of trichloroethylene vapor at sub-parts-per-billion concentrations in complex mixtures. Anal. Chem. 83(18):7198-7206.

Kim, S.K., D.R. Burris, H. Chang, J. Bryant-Genevier, and E.T. Zellers. 2012a. Microfabricated gas chromatograph for on-site determinations of trichloroethylene in indoor air arising from vapor intrusion. 1. Field evaluation. Environ.Sci. Technol. in press. 46(11):6065-6072.

Kim, S.K., D.R. Burris, J. Bryant-Genevier, K.A. Gorder, E.M. Dettenmair, and E.T. Zellers. 2012b. Microfabricated gas chromatograph for on-site determinations of TCE in indoor air arising from vapor intrusion. 2. Spatial/temporal monitoring. Environ. Sci. Technol. 46(11):6073-6080.

Klaper, R., B.J. Carter, C.A. Richter, P.E. Drevnick, M.B. Sandheinrich, and D.E. Tillitt. 2010. Corrigendum: Use of a 15 k gene microarray to determine gene expression changes in response to acute and chronic methylmercury exposure in the fathead minnow *Pimephales promelas* Rafinesque (72(9):2207- 2008). J. Fish Biol. 77(1):310.

Kolok, A.S., H.L. Schoenfuss, C.R. Propper, and T.L. Vail. 2011. Empowering citizen scientists: The strength of many in monitoring biologically active environmental contaminants. BioScience 61(8):626-630.

Kopecky, K.J., S. Davis, T.E. Hamilton, M.S. Saporito, and L.E. Onstad. 2004. Estimation of thyroid radiation doses for the Hanford thyroid disease study: Results and implications for statistical power of the epidemiological analyses. Health Phys. 87(1):15-32.

Korrick, S.A., L.M. Altshul, P.E. Tolbert, V.W. Burse, L.L. Needham, and R.R. Monson. 2000. Measurement of PCBs, DDE, and hexachlorobenzene in cord blood from infants born in towns adjacent to a PCB-contaminated waste site. J. Expo. Anal. Environ. Epidemiol. 10(6 Pt 2):743-754.

Kratz, T.K., P. Arzberger, B.J. Benson, C.Y. Chiu, K. Chiu, L. Ding, T. Fountain, D. Hamilton, P.C. Hanson, Y.H. Hu, F.P. Lin, D.F. McMullen, S. Tilak, and C. Wu. 2006. Towards a Global Lake Ecological Observatory Network. Publications of the Karelian Institute 145:51-63 [online]. Available: http://www.gleon.org/Gleon _Kratz_etal_2006.pdf [accessed Jan. 17, 2012].

Krause, A.R., C. Van Neste, L.R. Senesac, T. Thundat, and E. Finot. 2008. Trace explosive detection using photothermal deflection spectroscopy. J. App. Phys. 103(9):094906.

Lahr, J., and L. Kooistra. 2010. Environmental risk mapping of pollutants: State of the art and communication aspects. Sci. Total Environ. 408(18):3899-3907.

Lammel, G. 2004. Effects of time-averaging climate parameters on predicted multicompartmental fate of pesticides and POPs. Environ. Pollut. 128(1-2):291-302.

Lavrova, O.Y., and A.G. Kostianoy. 2011. Catastrophic oil spill in the Gulf of Mexico in April-May 2010. Izv. Atmos. Ocean. Phys. 47(9):1114-1118.

Leamer, E.E. 1978. Specification Searches: Ad Hoc Inference with Nonexperimental Data. New York: Wiley.

Lee, H.J., Y. Liu, B.A. Coull, J. Schwartz, and P. Koutrakis. 2011. A novel calibration approach of MODIS AOD data to predict $PM_{2.5}$ concentrations. Atmos. Chem. Phys. 11(5):7991-8002.

Lewis, P.R., P. Manginell, D.R. Adkins, R.J. Kottenstette, D.R. Wheeler, S.S. Sokolowski, D.E. Trudell, J.E. Byrnes, M. Okandan, J.M. Bauer, R.G. Manley, and C. Frye-Mason. 2006. Recent advancements in the gas-phase MicroChemLab. IEEE Sensors Journal 6(3):784-795.

Leyk, S., C.R. Binder, and J.R. Nuckols. 2009. Spatial modeling of personalized exposure dynamics: The case of pesticide use in small-scale agricultural production landscapes of the developing world. Int. J. Health Geo. 8:17.

Li, M., H.X. Tang, and M.L. Roukes. 2007. Ultra-senstive NEMS-based cantilever array for sensing, scanned probes, and very high-frequency applications. Nature Nanotech. 2(2):114-120.

Litton, C.D., K.D. Smith, R. Edwards, and T. Allen. 2004. Combined optical and ionization measurement techniques for inexpensive characterization of micrometer and submicrometer aerosols. Aerosol Sci. Technol. 38(11):1054-1062.

Longley, P., M. Goodchild, D. Maguire, and D. Rhind. 2005. Geographic Information Systems and Science. New York: Wiley.

Lyapustin, A., J. Martonchik, Y. Wang, I. Laszlo, and S. Korkin. 2011a. Multi-angle implementation of atmospheric correction (MAIAC): Part 1. Radiative transfer basis and look-up tables. J. Geophys. Res. 116: D03210, doi:10.1029/2010JD014985.

Lyapustin, A., Y. Wang, I. Laszlo, R. Kahn, S. Korkin, L. Remer, R. Levy, and J.S. Reid. 2011b. Multi-angle implementation of atmospheric correction (MAIAC): Part 2. Aerosol algorithm. J. Geophys. Res. 116: D03211, doi:10.1029/2010JD014986.

Maclachlan, J.C., M. Jerrett, T. Abernathy, M. Sears, and M.J. Bunch. 2007. Mapping health on the internet: A new tool for environmental justice and public health research. Health Place 13(1):72-86.

MacLeod, M., D. Woodfine, J. Brimacombe, L. Toose, and D. Mackay. 2002. A dynamic mass budget for toxaphene in North America. Environ. Toxicol. Chem. 21(8):1628-1637.

MacLeod, M., M. Scheringer, T.E. McKone, and K. Hungerbühler. 2010. The state of multimedia mass-balance modeling in environmental science and decision making. Environ. Sci. Technol. 44(22):8360-8364.

Maina, J., V. Venus, M.R. McClanahan, and M. Ateweberhan. 2008. Modeling susceptibility of coral reefs to environmental stress using remote sensing data and GIS models. Ecol. Model. 212(3-4):180-199.

Malley, D.F., K.N. Hunter, and G.R. Webster. 1999. Analysis of diesel fuel contamination in soils by near-infrared reflectance spectrometry and solid phase microextraction-gas chromatography. Soil Sediment Contam. 8(4):481-489.

Mattingly, C.J. 2009. Chemical databases for environmental health and clinical research. Toxicol. Lett. 186(1):62-65.

Mattingly, C.J., T.E. McKone, M.A. Callahan, J.A. Blake, and E.A. Cohen-Hubal. 2012. Providing the missing link: The exposure science ontology ExO. Environ. Sci. Technol. 46(6):3046-3053.

Maxwell, S.K., J.R. Meliker, and P. Goovaerts. 2010. Use of land surface remotely sensed satellite and airborne data for environmental exposure assessment in cancer research. J. Expo. Sci. Environ. Epidemiol. 20(2):176-185.

McCreanor, J., P. Cullinan, M.J. Nieuwenhuijsen, J. Stewart-Evans, E. Malliarou, L. Jarup, R. Harrington, I.K. Svartengren, P. Ohman-Strickland, K.F. Chung, and J. Zhang. 2007. Respiratory effects of exposure to diesel traffic persons with asthma. N. Engl. J. Med. 357(23):2348-2358.

McKone, T.E., and M. MacLeod. 2003. Tracking multiple pathways of human exposure to persistent multimedia pollutants: Regional, continental and global-scale models. Annu. Rev. Environ. Resour. 28:463-492.

McKone, T.E., R. Castorina, M.E. Harnly, Y. Kuwabara, B. Eskenazi, and A. Bradman. 2007. Merging models and biomonitoring data to characterize sources and pathways of human exposure to organophosphorous pesticides in the Salinas Valley of California. Environ. Sci. Technol. 41(9):3233-3240.

Mishra, D., C. Cho, S. Ghosh, A. Fox, C. Downs, P. Merani, P. Kirui, N. Jackson, and S. Mishra. 2012. Post-spill state of the marsh: Remote estimation of the ecological impact of the Gulf of Mexico oil spill on Louisiana Salt Marshes. Remote Sens. Environ. 118:176-185.

Molitor, J., M. Jerrett, C.C. Chang, N.T. Molitor, J. Gauderman, K. Berhane, R. McConnell, F. Lurmann, J. Wu, A. Winer, and D. Thomas. 2007. Assessing uncertainty in spatial exposure models for air pollution health effects assessment. Environ. Health Perspect. 115(8):1147-1153.

Morland, K.B., and K.R. Evenson. 2009. Obesity prevalence and the local food environment. Health Place 15(2):491-495.

Mun, M., S. Reddy, K. Shilton, N. Yau, J. Burke, D. Estrin, M. Hansen, E. Howard, R. West, and P. Boda. 2009. PEIR, the personal environmental impact report, as a platform for participatory sensing system research. Pp. 55-68 in Proceedings of the

7th Annual International Conference on Mobile Systems, Applications and Services-MobiSys '09, June 22-25, 2009, Krakow, Poland. New York: ACM.

NASA (National Aeronautics and Space Administration). 2011. HyspIRI Mission Study. Jet Propulsion Laboratory. California Institute of Technology [online]. Available: http://hyspiri.jpl.nasa.gov/ [accessed Oct. 28, 2011].

Nebert, D.W., T.P. Dalton, A.B. Okey, and F.J. Gonzalez. 2004. Role of aryl hydrocarbon receptor-mediated induction of the CYP1 enzymes in environmental toxicity and cancer. J. Biol. Chem. 279(23):23847-23850.

Nicholson, J.K., and J.C. Lindon. 2008. Systems biology: Metabonomics. Nature 455(7216):1054-1056.

Noy, N.F., and D.L. McGuinness. 2001. Ontology Development101: A Guide to Creating Your First Ontology. KSL-01-05. Knowledge System Laboratory, Stanford University, CA [online]. Available: http://www.ksl.stanford.edu/KSL_Abstracts/KSL-01-05.html [accessed May 31, 2012].

NRC (National Research Council). 2007. Models in Environmental Regulatory Decision Making. Washington, DC: National Academies Press.

NRC (National Research Council). 2009. Science and Decisions: Advancing Risk Assessment. Washington, DC: National Academies Press.

OBO Foundry (The Open Biological and Biomedical Ontologies Foundry). 2012. Exposure Ontology [online]. Available: http://obolibrary.org/cgi-bin/detail.cgi?id=exo [accessed June 4, 2012].

Odermatt, D., A. Gitelson, V.E. Brando, and M. Schaepman. 2012. Review of constituent retrieval in optically deep and complex waters from satellite imagery. Remote Sens. Environ. 118:116-126.

Offit, K. 2011. Personalized medicine: New genomics, old lessons. Hum. Genet. 130(1):3-14.

Ong, C.H., T.J. Cudahy, M.S. Caccetta, and M.S. Piggott. 2003. Deriving quantitative dust measurements related to iron ore handling from airborne hyperspectral data. Min. Technol. IMM Trans. Sect. A 112(3):158-163.

Pastorello, G.Z., G.A. Sanchez-Azofeifa, and M.A. Nascimento. 2011. Enviro-net: From networks of ground-based sensor systems to a web platform for sensor data management. Sensors 11(6):6454-6479.

Patel, C.J., J. Bhattacharya, and A.J. Butte. 2010. An Environment-Wide Association Study (EWAS) on Type 2 Diabetes Mellitus. PLoS One 5(5):e10746.

Patel, S.V., T.E. Mlsna, B. Fruhberger, E. Klaassen, S. Cemalovic, and D.R. Baselt. 2003. Chemicapactive microsensors for volatile organic compound detection. Sensor. Actuat. B Chem. 96(3):541-553.

Paulos, E., R.J. Honicky, and E. Goodman. 2007. Sensing atmosphere. The 5th ACM Conference on Embedded Network Sensor Systems-AMC SenSys, November 6-9, 2007, Sydney, Australia [online]. Available: http://www.paulos.net/papers/2007/Sensing%20Atmosphere%20(Sensys%202007%20Workshop).pdf [accessed Jan. 18, 2012].

Pelletier, B., R. Santer, and J. Vidot. 2007. Retrieving of particulate matter from optical measurements: A semiparametric approach. J. Geophys. Res. Atmos. 112:D06208, doi:10.1029/2005JD006737.

Peters, A., G. Hoek, and K. Katsouyanni. 2012. Understanding the link between environmental exposures and health: Does the exposome promise too much? J. Epidemiol. Community Health 66(2):103-105.

Richardson, D.B., and A. Ciampi. 2003. Effects of exposure measurement error when an exposure variable is constrained by a lower limit. Am. J. Epidemiol. 157(4):355-363.

Roberts, A.P., J.T. Oris, G.A. Burton, and W.H. Clements. 2005. Gene expression in caged fish as a first-tier indicator of contaminant exposure in streams. Environ. Toxicol. Chem. 24(12):3092-3098.

RTI International. 2008. New Technology Used to Increase Accuracy, Ease Measurement of Harmful Environmental Exposure. RTI International News: September 16, 2008 [online]. Available: http://www.rti.org/page.cfm?objectid=4749BFB4-CCC0-2F2 C-9D8D787BCE30A49D [accessed Jan. 11, 2012].

Rundel, P.W., E.A. Graham, M.F. Allen, J.C. Fisher, and T.C. Harmon. 2009. Environmental sensor networks in ecological research. New Phytol. 182(3):589-607.

Sanchez, Y.A., K. Deener, E. Cohen Hubal, K. Knowlton, D. Reif, and D. Segal. 2010. Research needs for community-based risk assessment: Findings from multidisciplinary workshop. J. Expo. Sci. Environ. Epidemiol. 20(2):186-195.

Sanchez, W., W. Sremski, B. Piccini, O. Palluel, E. Maillot-Marechal, S. Betoulle, A. Jaffal, S. Ait-Aissa, F. Brion, E. Thybaud, N. Hinfray, and J.M. Porcher. 2011. Adverse effects in wild fish living downstream from pharmaceutical manufacture discharges. Environ. Int. 37(8):1342-1348.

Sarangapani, R., J. Teeguarden, K.P. Plotzke, J.M. McKim, Jr, and M.E. Andersen. 2002. Dose-response modeling of cytochrome p450 induction in rats by octamethylcyclotetrasiloxane. Toxicol. Sci. 67(2):159-172.

Schenker, U., M. MacLeod, M. Scheringer, and K. Hungerbühler. 2005. Improving data quality for environmental fate models: A least-squares adjustment procedure for harmonizing physicochemical properties of organic compounds. Environ. Sci. Technol. 39(21):8434-8441.

Scheringer, M., F. Wegmann, K. Fenner, and K. Hungerbühler. 2000. Investigation of the cold condensation of persistent organic pollutants with a global multimedia fate model. Environ. Sci. Technol. 34(9):1842-1850.

Scheringer, M., M. Stroebe, F. Wania, F. Wegmann, and K. Hungerbühler. 2004. The effect of export to the deep sea on the long-range transport potential of persistent organic pollutants. Environ. Sci. Pollut. Res. Int. 11(1):41-48.

Scholz, S., and I. Mayer. 2008. Molecular biomarkers of endocrine disruption in small model fish. Mol. Cell. Endocrinol. 293(1-2):57-70.

Schwartz, D., and F. Collins. 2007. Environmental biology and human disease. Science 316(5825):695-696.

Senesac, L., and T.G. Thundat. 2008. Nanosensors for trace explosive detection. Materials Today 11(3):28-36.

Seto, E., E. Martin, A. Yang, P. Yan, R. Gravina, I. Lin, C. Wang, M. Roy, V. Shia, and R. Bajcsy. 2010. Opportunistic Strategies for Lightweight Signal Processing for Body Sensor Networks. Proceedings of the 3rd International Conference on Pervasive Technology Related to Assistive Environments-PETRA, June 23-25, 2010, Samos, Greece [online]. Available: http://www.eecs.berkeley.edu/~yang/paper/Se toPETRAE2010.pdf [accessed Jan. 12, 2012].

Seto, E., P. Yan, P. Kuryloski, R. Bajcsy, T. Abresch, E. Henricson, and J. Han. 2011. Mobile Phones as Personal Environmental Sensing Platforms: Development of the CalFit Systems. Abstract S-0034 in Abstracts of the 23rd Annual Conference of the International Society of Environmental Epidemiology (ISEE), September 13 - 16, 2011, Barcelona, Spain [online]. Available: http://ehp03.niehs.nih.gov/article/

fetchArticle.action?articleURI=info%3Adoi%2F10.1289%2Fehp.isee2011 [accessed Sept. 4, 2012].

Shalat, S.L., A.A. Stambler, Z. Wang, G. Mainelis, O.H. Emoekpere, M. Hernandez, P.J. Lioy, and K. Black. 2011. Development and in-home testing of the Pretoddler Inhalable Particulate Environmental Robotic (PIPER Mk IV) sampler. Environ. Sci. Technol. 45(7):2945-2950.

Shelley, S. 2008. Update. Nanosensors: Evolution, not revolution…yet. CEP 104(6):8-12.

Shi, Q., H. Hong, J. Senior, and W. Tong. 2010. Biomarkers for drug-induced liver injury. Expert Rev. Gastroenterol. Hepatol. 4(2):225-234.

Shipler, D.B., B.A. Napier, W.T. Farris, and M.D. Freshley. 1996. Hanford environmental dose reconstruction project–an overview. Health Phys. 71(4):532-544.

Short, N.M., Sr. 2011. Introduction: Technical and Historical Perspectives of Remote Sensing. Remote Sensing Tutorial [online]. Available: http://rst.gsfc.nasa.gov/Intro/Part2_1.html [accessed Jan. 18, 2012].

Shoval, N., and M. Isaacson. 2006. Application of tracking technologies to the study of pedestrian spatial behavior. Prof. Geog. 58(2):172-183.

Simon, S.L., J.E. Till, R.D. Lloyd, R.L. Kerber, D.C. Thomas, S. Preston-Martin, J.L. Lyon, and W. Stevens. 1995. The Utah leukemia case-control study: Dosimetry methodology and results. Health Phys. 68(4):460-471.

Smith, P.N., G.P Cobb., C. Godard-Codding, D. Hoff, S.T. McMurry, T.R. Rainwater, and K.D. Reynolds. 2007. Contaminant exposure in terrestrial vertebrates. Environ. Pollut. 150(1):41-64.

Soltow, Q.A., F.H. Strobel, K.G. Mansfield, L. Wachtman, Y. Park, and D.P. Jones. In press. High-performance metabolic profiling with dual chromatography-Fourier-transform mass spectrometry (DC-FTMS) for study of the exposome. Metabolomics in press.

Stahl, R.G., T.S. Bingman, A. Guiseppi-Elie, and R.A. Hoke. 2010. What biomonitoring can and cannot tell us about causality in human health and ecological risk assessments. Hum. Ecol. Risk Assess. 16(1):74-86.

Steenland, K., and D. Savitz. 1997. Topics in Environmental Epidemiology. New York: Oxford University Press.

Stram, D.O., and K.J. Kopecky. 2003. Power and uncertainty analysis of epidemiological studies of radiation-related disease risk in which dose estimates are based on a complex dosimetry system: Some observations. Radiat. Res. 160(4):408-417.

Su, J.G., M. Winters, M. Nunes, and M. Brauer. 2010. Designing a route planner to facilitate and promote cycling in Metro Vancouver, Canada. Trans. Res. Part A 44(7):495-505.

Swayze, G.A., R.N. Clark, S.J. Sutley, T.M. Hoefen, G.S. Plumlee, G.P. Meeker, I.K. Brownfield, K.E. Livo, and L.C. Morath. 2006. Spectroscopic and x-ray diffraction analyses of asbestos in the World Trade Center dust: Asbestos content of the settled dust. Pp. 40-65 in Urban Aerosols and Their Impact: Lessons Learned from the World Trade Center Tragedy, J.S. Gaffney, and N.A. Marley, eds. American Chemical Society Symposium Series 919. Oxford: Oxford University Press.

Tan, Y.M., K.H. Liao, and H.J. Clewell, III. 2007. Reverse dosimetry: Interpreting trihalomethanes biomonitoring data using physiologically based pharmacokinetic modeling. J. Expo. Sci. Environ. Epidemiol. 17(7):591-603.

Tang, Z., H. Wu, D. Du, J. Wang, H. Wang, W.J. Qian, D.J. Bigelow, J.G. Pounds, R.D. Smith, and Y. Lin. 2010. Sensitive immunoassays of nitrated fibrinogen in human biofluids. Talanta 81(4-5):1662-1669.

Teeguarden, J.G., P.J. Deisinger, T.S. Poet, J.C. English, W.D. Faber, H.A. Barton, R.A. Corley, and H.J. Clewell, III. 2005. Derivation of a human equivalent concentration for n-butanol using a physiologically based pharmacokinetic model for n-butyl acetate and metabolites n-butanol and n-butyric acid. Toxicol. Sci. 85(1):429-446.

Thomas, D.C., D. Stram, and J. Dwyer. 1993. Exposure measurement error: Influence on exposure-disease. Relationships and methods of correction. Annu. Rev. Public Health 14:69-93.

Timchalk, C., J.A. Campbell, G. Liu, Y. Lin, and A.A. Kousba. 2007. Development of a non-invasive biomonitoring approach to determine exposure to the organophosphorus insecticide chlorpyrifos in rat saliva. Toxicol. Appl. Pharmacol. 219(2-3):217-225.

Todaka, T., H. Hirakawa, J. Kajiwara, T. Hori, K. Tobiishi, D. Yasutake, D. Onozuka, S. Sasaki, C. Miyashita, E. Yoshioka, M. Yuasa, R. Kishi, T. Iida, and M. Furue. 2010. Relationship between the concentrations of polychlorinated dibenzo-p-dioxins, polychlorinated dibenzofurans, and polychlorinated biphenyls in maternal blood and those in breast milk. Chemosphere 78(2):185-192.

van der Meer, F.D., P.M. van Dijk, H. van der Werrf, and H. Yang. 2002. Remote sensing and petroleum seepage: A review and case study. Terra Nova 14(1):1-17.

van Donkelaar, A., R.V. Martin, M. Brauer, R. Kahn, R. Levy, C. Verduzco, and P.J. Villeneuve. 2010. Global estimates of ambient fine particulate matter concentrations from satellite-based aerosol optical depth: Development and application. Environ. Health Perspect. 118(6):847-855.

Vazquez-Prokopec, G.M., S.T. Stoddard, V. Paz-Soldan, A.C. Morrison, J.P. Elder, T.J. Kochel, T.W. Scott, and U. Kitron. 2009. Usefulness of commercially available GPS data-loggers for tracking human movement and exposure to dengue virus. Int. J. Health Geo. 8:68.

Villenueuve, D.L., and N. Garcia-Reyero. 2011. Vision and strategy: Predictive ecotoxicology in the 21st century. Environ. Toxicol. Chem. 30(1):1-8.

Wacholder, S., D.T. Silverman, J.K. McLaughlin, and J.S. Mandel. 1992. Selection of control in case-control studies. III. Design options. Am. J. Epidemiol. 135(9):1042-1050.

Walt, D.R. 2005. Electronic noses: Wake up and smell the coffee. Anal. Chem. 77(3):45A.

Wang, Z., S.L. Shalat, K. Black, P.J. Lioy, A.A. Stambler, O.H. Emoekpere, M. Hernandez, T. Han, M. Ramagopal, and G. Mainelis. 2012. Use of a robotic sampling platform to assess young children's exposure to indoor bioaerosols. Indoor Air 22(2):159-169.

Wania, F., and D. Mackay. 1993. Global fractionation and cold condensation of low volatility organochlorine compounds in polar regions. Ambio 22(1):10-18.

Wania, F., and D. Mackay. 1999. The evolution of mass balance models of persistent pollutant fate in the environment. Environ. Pollut.100(1-3):223-240.

Wania, F., and Y. Su. 2004. Quantifying the global fractionation of polychlorinated biphenyls. Ambio 33(3):161-168.

Ward, M.H., J.R. Nuckols, S.J. Weigel, S.K. Maxwell, K.P. Cantor, and R.S. Miller. 2000. Identifying populations potentially exposed to agricultural pesticides using remote sensing and a geographic information system. Environ. Health Perspect. 108(1):5-12.

Ward, M.H., J. Lubin, J. Giglierano, J.S. Colt, C. Wolter, N. Bekiroglu, D. Camann, P. Hartge, and J.R. Nuckols. 2006. Proximity to crops and residential exposure to agricultural herbicides in Iowa. Environ. Health Perspect. 114(6):893-897.

Whitehead, A., B. Dubansky, C. Bodinier, T.I. Garcia, S. Miles, C. Pilley, V. Raghunathan, J.L. Roach, N. Walker, R.B. Walter, C.D. Rice, and F. Galvez. 2011. Genomic and physiological footprint of the Deepwater Horizon oil spill on resident marsh fishes. Proc. Natl. Acad. Sci. USA [online]. Available: http://www.pnas.org/content/early/2011/09/21/1109545108.full.pdf [accessed Feb. 16, 2012].

Wild, C.P. 2005. Complementing the genome with an "exposome": The outstanding challenge of environmental exposure measurement in molecular epidemiology. Cancer Epidemiol. Biomarkers Prev. 14(8):1847-1850.

Williamson, C.E., J.E. Saros, and D.W. Schindler. 2009. Sentinels of change. Science 323(5916): 887-888.

Wilson, M.L. 2002. Emerging and vector-borne diseases: Role of high spatial resolution and hyperspectral images in analyses and forecasts. J. Geograph. Syst. 4(1):31-42.

Winkelmann, K.H. 2005. On the Applicability of Imaging Spectroscopy for the Detection and Investigation of Contaminated Sites with Particular Consideration Given to the Detection of Fuel Hydrocarbon Contamination in Soil. Ph.D. Dissertation, Brandenburgischen Technischen Universität, Cottbus.

Wootton, R., and L. Bonnardot. 2010. In what circumstances is telemedicine appropriate in the developing world? JRSM Short Rep. 1(5):37.

Wu, Y.Z., J. Chen, J.F. Ji, Q.J. Tian, and X.M. Wu. 2005. Feasibility of reflectance spectroscopy for the assessment of soil mercury contamination. Environ. Sci. Technol. 39(3):873-878.

Yuan, Y.M., W. Xiong, Y.H. Fang, T.G. Lan, and D.C. Li. 2010. Detection of oil spills on water by differential polarization FTIR spectrometry [in Chinese]. Guang Pu Xue Yu Guang Pu Fen Xi 30(8): 2129-2132.

Zartarian, V.G., H. Ozkaynak, J.M. Burke, M.J. Zufall, M.L. Rigas, and E.J. Furtaw, Jr. 2000. A modeling framework for estimating children's residential exposure and dose to chlorpyrifos via dermal residue contact and nondietary ingestion. Environ Health Perspect. 108(6):505-514.

Zhang, X., M.E. Monroe, B. Chen, M.H. Chin, T.H. Heibeck, A.A. Schepmoes, F. Yang, B.O. Petritis, D.G. Camp II, J.G. Pounds, J.M. Jacobs, D. J. Smith, D. J. Bigelow, R.D. Smith, and W. Qian. 2010. Endogenous 3,4-dihydroxyphenylalanine and dopaquinone modifications on protein tyrosine: Links to mitochondrially derived oxidative stress via hydroxyl radical. Mol. Cell Proteomics 9(6):1199-1208.

6

Promoting and Sustaining Public Trust in Exposure Science

The vision of exposure science articulated in the preceding chapters requires broad public support for gathering information on human and environmental exposures. That support involves understanding the aims of exposure science and a perspective that exposure science is integral to decision-making, including developing mitigation and prevention strategies that advance societal goals and objectives. If support is to weather the inevitable challenges, it is critical that researchers and stakeholders share values of and expectations for exposure science. In addition, appropriate safeguards and definable ethical standards to protect volunteers who give their time for exposure research are needed. Those considerations illustrate that the long-term success of human and environmental exposure science requires that attention be focused on the ethical and societal issues raised by modern research practice.

Although human exposure studies typically present only minor risks to research volunteers, several studies have sparked controversy. For example, the Children's Environment Exposure Research Study (CHEERS), a proposed study that was intended to examine how children may be exposed to pesticides and other household chemicals generated considerable mass-media interest and contributed to fears that children would be placed at increased risk. The concerns prompted the U.S. Environmental Protection Agency to cancel its plans for the study (Resnik and Wing 2007). Reactions to the proposed study resulted in raising the bar for the conduct of observational exposure studies and led to the publication of *Scientific and Ethical Approaches for Observational Exposure Studies* (EPA 2008), which provides guidance for conducting observational studies. The CHEERS study, together with other controversial studies, highlight a number of ethical and societal issues that need to be navigated successfully to achieve the committee's vision for human and environmental exposure science.

In this chapter, the committee addresses considerations related to the promotion of public trust in exposure science. Although none of the issues discussed is peculiar to exposure science, how they present themselves can pose special challenges for observational studies. That is particularly true of ethical issues associated with ubiquitous technologies—new personal and environmental monitoring tools—whose use and availability are expected to become more prevalent. As technologies evolve our capacity to detect chemicals and other stressors at increasingly lower levels is outpacing our capacity to interpret what the findings mean for health, subclinical effects, or even for elucidating potential exposure reduction strategies, creating scientific and ethical challenges for society.

PROTECTING RESEARCH VOLUNTEERS

Exposure-science research may involve observational studies of humans and the environments in which they live. The ability to conduct exposure studies in all populations, particularly among the most vulnerable (for example, the elderly, children, and the infirm) is critical to understanding and preventing harmful exposures and risks in society. As with all research involving human subjects, it is vital that measures be implemented at the beginning of a study to protect the rights and welfare of participants (volunteers). In cases that lack such protections, research subjects may be needlessly placed at increased risk, which can threaten not only their well being but the long-term success of human and environmental exposure-science research.

All exposure-science research involving human subjects must be reviewed and approved by an appropriate institutional review board (IRB) and be consistent with national ethics standards. Many of the ethical considerations that should guide the conduct of observational exposure studies are discussed in the recent EPA report *Scientific and Ethical Approaches for Observational Exposure Studies*, which provides well-defined and documented procedures for conducting exposure-related research. That report describes two primary elements for human-subject protections: review by an impartial IRB and informed consent of research subjects.

High standards of ethical conduct in human-subjects research are enforced by multiple oversight committees or review boards, which examine both scientific and ethical details of the proposed research. IRBs include appropriate medical and scientific experts who are not directly involved in the conduct of the research and who do not have any personal interests in the study. They also have members outside the scientific community. In deciding whether a research study is acceptable, IRB members are asked to consider whether or not the foreseeable risks to participants are appropriately minimized and acceptable in light of the potential benefits to participants and to society.

The process of seeking informed consent of participants is a second key element of ethical conduct in research. Informed consent ensures that subjects give their voluntary and informed consent to be participants in a research study.

In addition to consenting to participate in a study, participants can consent on whether or not they want their results (Morello-Frosch et al. 2009). (Additional discussion is found in the section, Returning Research Results.) Researchers conducting observational exposure studies are expected to seek permission from research subjects who must be informed about the goals of the research, the specific procedures involved, and the potential benefits of and risks posed by participation. That kind of communication does not end when a research subject signs a permission form or otherwise indicates his or her willingness to participate. As a study progresses and new findings are discovered, investigators are expected to communicate to research subjects any changes that have been made in the study protocol. It is through such regular communication that consent to participate is reaffirmed. If an important risk is detected, the investigators are also obliged to report it to the participants and the funding agency involved.

Typically, observational studies that do not involve efforts to mitigate human exposures present minor risks to the people who participate as research subjects. However, in addition to minor burdens associated with the collection of exposure data, without full communication such studies may also increase participants' exposure in unintended ways, for example, by promoting an unwarranted sense of protection to subjects through their wearing of exposure-monitoring devices. Observational studies can also present risks related to the unintended disclosure of private information; for example, release of medical records might impair a person's ability to obtain employment. To address those possible risks, investigators should strive to ensure that data are protected and that research subjects understand any potential risks associated with their participation.

The committee acknowledges in addition to observational exposure studies, there are intentional dosing studies that can involve potential increases in harmful exposures. For example, a study conducted in London, England, had asthmatic volunteers walk routes with low and high exposures to air pollution (McCreanor et al. 2007). In another study, men suffering from congestive heart failure were exposed to high concentrations of diesel exhaust, which has been linked to an increased risk of a coronary event (Mills et al. 2007). Such studies raise additional ethical considerations (see NRC 2004); but these studies fall outside the scope of this committee's charge.

It is critical that scientists conducting research with human subjects not have conflicts of interest that may impair their ability to conduct the research ethically. Financial relationships with industry sponsors of research should be disclosed to IRBs and other research-oversight boards, such as university committees on conflict of interest. Sources of research funding should be disclosed to potential volunteers when they are approached to participate in a study, as should any financial relationships of the investigators with commercial entities that may be affected by the outcome of the research. Such disclosures can promote public trust by increasing transparency and giving study volunteers the opportunity to assess the full array of potential benefits and risks associated with their participation in the research.

PROMOTING PUBLIC TRUST

Public trust must be promoted to foster support for exposure-science research. Mass-media coverage of higher-visibility studies has at times mischaracterized key aspects of study design, reinforcing distrust of regulatory agencies and fueling concerns about the identification of human and environmental stressors. To maintain public confidence in the integrity of exposure science, innovative forms of public engagement are required. They should foster public understanding of the purpose, benefits, and limitations of human and environmental exposure-science research and its potential contributions to public health.

Engaging the public through education may be needed to improve understanding of how exposure-science studies have made great strides in identifying and reducing harmful exposures. Examples are described in earlier chapters and in a recently published series of exposure-science digests (Graham 2010). These and other exposure-science studies have contributed to greater understanding of the health effects of hazardous environments and have helped to reduce the likelihood of future adverse exposures.

Exposure scientists should engage members of the public in identifying and addressing relevant health concerns. To advance that goal, it is critical that members of the public be able to access information from human and environmental exposure studies. Although exposure studies are often intended to serve as the basis of regulatory policy, the data that they produce may have relevance for other purposes. In particular, members of the public may find results of exposure studies useful in addressing local environmental conditions—for example, in modifying local planning or zoning ordinances or blocking construction of new facilities on the basis of data showing that the community will be burdened by increased exposure to environmental stressors. Exposure information may also help members of the public in making personal choices about their environments, including their behaviors, diet, housing, employment, or the purchase of personal products, and can be used by community organizations and individuals to provide input to decisionmakers.

COMMUNITY ENGAGEMENT AND
STAKEHOLDER PARTICIPATION

Exposure science can help communities to identify and address differential, cumulative, and emergent exposures. Community members can be among the first to identify an exposure of concern. Once the design of a community-based exposure study begins, community members may be found to have a wealth of information about the exposures of potential concern. Communities are all too often the "subjects of" health studies and have little input into the problems to be studied. Recognizing potential community knowledge and adopting a spirit of openness in disseminating exposure-related data are pivotal for building trust. Correspondingly, conducting exposure studies in the absence of community input or failing to maintain communication with affected communities may greatly diminish public confidence in exposure science and reinforce distrust of scientists engaged in this work (NRC 1989; 1996). Engaging members of differentially affected communities is critical for increasing public trust and improving stakeholder participation in exposure science. Federal and private funding agencies have been providing increased support for community-engaged exposure studies, recognizing that this contributes to improving study results through better recruitment and retention of study participants and through dissemination of study results to diverse public audiences (Brown et al. 2012).

Open discussion is necessary if community members are to have input into problem formulation and ultimately to benefit from research findings. The groundwork for such discussion lies in effective multidirectional communication among all stakeholders, including members of affected communities, scientists, public-health agencies, and policy-makers (Chess et al. 1988; NRC 1989, 1996; Lundgren and McMakin 2009). Effective communication underlies all stages of research, from identifying the stakeholders who should be at the table to understanding how research findings can best be shared with community members. Outreach and engagement not only provide the vehicle for disseminating the results of exposure studies but provide the opportunity for researchers to listen to community members so that they can identify preferred approaches and formats for dissemination. Such information may assist communities in developing new public-health interventions, environmental policies, or community-development initiatives to reduce harmful exposures.

USE OF COMMUNITY-BASED PARTICIPATORY RESEARCH

A trend in public-health research is the engagement of affected communities and populations through the use of community-based participatory research (CBPR). Members of differentially burdened and exposed communities have turned to CBPR as an approach to communicate their environmental-health concerns to scientists and promote collaborative research. As part of CBPR, community-based organizations use their grassroots activism and resources, "expert local knowledge", and university partners to develop a framework for studying differential health effects at the local level (Wilson et al. 2008; Baron and Wilson 2011). Critical to this approach is an effort to put scientific knowledge into practice by dissolving traditional boundaries between knowledge and action. CBPR acknowledges the community as a unit of identity and builds on community strengths and resources to facilitate more equitable partnerships that involve power-sharing and community empowerment (Israel et al. 1998).

Many community-based environmental justice and health organizations exist across the United States, for example, West Harlem Environmental Action, in New York City; Concerned Citizens of Tillery, North Carolina; Alternatives for Community and Environment, based in Boston, Massachusetts; and the Environmental Health Coalition, in San Diego. Many of these organizations have engaged in community-driven research to obtain locally relevant exposure data on a variety of environmental health issues, including air pollution in metropolitan areas, housing stock, burden of pollution from industries, locally unwanted land uses, transportation issues, and industrial animal production (Shepard et al. 2002). Using CBPR, they and other organizations have leveraged their local knowledge in collaboration with university partners to develop a framework for addressing environmental health and justice issues at the local level (Wilson et al. 2008). Using that approach, community-based organizations have become more involved in creating research agendas that advance the goals of differentially affected communities (see for example, Box 6-1 which describes the role of CBPR in reducing adverse exposures and advancing the health of the community of Spartansburg, SC).

BOX 6-1 Case Study of Exposure Justice and Community Engagement: ReGenesis in Spartanburg, SC

The city of Spartanburg is in northwestern South Carolina and has a population of 40,000, about 50% black and 50% white (EPA 2003, 2006). This former "textile town" has undergone a transformation with its revitalized downtown and a concentration of international business firms within the city limits (EPA 2003, 2006; Fleming 2004). However, the Arkwright and Forest Park neighborhoods, just beyond the city's downtown, are two predominantly black neighborhoods with a combined population of almost 5,000 residents that has not benefited from the revitalization efforts (EPA 2003, 2006; Fleming 2004). The closing of local mills and plants and the lack of zoning regulations and land-use controls (EPA 2003, 2006; Fleming 2004; Habisreutinger and Gunderson 2006) left the population poor (25% in poverty) and underemployed (10% unemployment) (EPA 2003, 2006).

The two neighborhoods are affected by environmental exposures. For example, the residents were exposed to a 40-acre International Mineral and Chemicals (IMC) fertilizer plant (a Superfund site); the Arkwright dump, a 30-acre former municipal landfill (a Superfund site); the Rhodia chemical plant (in operation); the Mt. Vernon textile mill (in operation); and six brownfields (EPA 2003, 2006; Fleming 2004; Habisreutinger and Gunderson 2006; ReGenesis 2008). About 4,700 people lived within 1 mile of the IMC site and 200 within 0.25 mile of the landfill (EPA 2003, 2006; Fleming 2004; ReGenesis 2008). Those exposures have been associated with a high rate of cancer—particularly bone, colon, and lung cancer—and high rates of respiratory illnesses, adult and infant mortality, miscarriages, and birth defects (EPA 2003, 2006; ReGenesis 2008). In addition, neighborhood residents had a poor transportation infrastructure, inadequate sewer and water services, lack of access to medical care, public-safety issues, few economic opportunities, and declining property values (EPA 2003, 2006).

(Continued)

> **BOX 6-1 Continued**
>
> In 1997, Harold Mitchell, a resident concerned about the environmental contamination in his community, began organizing community meetings to discuss environmental justice and health issues (EPA 2003, 2006; Fleming 2004; ReGenesis 2008). The meetings empowered local residents and motivated efforts by the government and industry to clean up the contaminated Superfund sites and brownfields. This community-driven collaboration became known as the "ReGenesis Project". ReGenesis built an environmental-justice partnership with the city of Spartanburg, the county, the EPA Region 4 Office of Environmental Justice, the Department of Health and Environmental Control, the Spartanburg Housing Authority, the county's Community and Economic Development Department, local industry, and the University of South Carolina to address environmental effects on local health and adopt strategies to revitalize the Arkwright and Forest Park neighborhoods (EPA 2003, 2006).
>
> With help from EPA Region 4, in 2000 ReGenesis was designated a national demonstration project of the Federal Interagency Working Group on Environmental Justice, giving it access to financial resources, technical experts, and information (EPA 2003, 2006). With that designation, new funding was made available, and local, state, and federal agencies began to understand that action was needed in the Arkwright and Forest Park neighborhoods to improve public health (EPA 2003, 2006). Spartanburg County was awarded additional funding through EPA's Brownfield Initiative to perform site assessments of the brownfields (EPA 2003, 2006; Habisreutinger and Gunderson 2006; ReGenesis 2008). The brownfields assessment found contamination and led to government agencies' providing additional funding to clean up the sites for redevelopment (EPA 2003, 2006; Habisreutinger and Gunderson 2006; ReGenesis 2008). The success of ReGenesis in working with its collaborative partners for assessment, cleanup, and redevelopment of brownfields and other industrial sites also led to additional efforts to improve the health-promoting infrastructure of the Arkwright and Forest Park neighborhoods.

The CBPR approach allows the research process to increase a community's ability to study differential and cumulative exposures, address environmental justice and health issues, and increase engagement of minority-group and low-income stakeholders (Minkler et al. 2006; Wilson et al. 2008; Baron and Wilson 2011). For example, CBPR was used to obtain community-generated dietary consumption data for subsistence anglers in the East River of New York. These data, together with fish tissue contaminant concentration data were used by EPA in calculating risk estimates for exposures to contaminants in fish (for example, cadmium, mercury, and dioxins) consumed by the local, ethnic populations, when no appropriate dietary assessment data were previously available for this high-risk group (Corburn 2002). Members of the community were essential for collecting this consumption data given the cultural and language barriers of the anglers. Similarly CBPR has been used to help assess farmworkers' pesticide exposures and associated health effects (Acury et al. 2001); a particularly difficult population to study given their transience and changing occupational status, language barriers, and questions regarding immigration status. Kamel et al. (2001) used community collaboration (working with the Farmworker Association of Florida and relying on the members' expertise) to identify and recruit a valid sample of farmworkers for an epidemiologic study.

CHALLENGES AHEAD

The new research paradigms and exposure-monitoring technologies described in the preceding chapters will magnify current debates about exposure science and pose additional challenges in maintaining public trust. The potential for increasing private-sector involvement in this field of research may raise additional concerns about personal privacy. Future studies may seek to relate environmental exposure data to other large databases, including data on cellular-telephone use, global positioning system location, food consumption, and consumer shopping patterns. Although it may not be possible to anticipate the full array of challenges that lie ahead, several important issues can be identified now.

Returning Research Results

A difficult challenge in exposure studies concerns whether, when, and how to return exposure information to study participants (Schulte and Singal 1996). That issue can be raised at both at an individual level and a community level and can be especially difficult in contexts in which exposure studies take a long time, there may not be a direct relationship between researchers and subjects, or if the exposure data do not provide clear insights regarding the potential health effects, sources, or pathways of exposure. Questions about the return of personal exposure information will become more common as new collection methods obtain "personalized" forms of exposure data. Such data might be collected with ubiquitous monitoring devices, such as personal cellular telephones (see discussion in Chapter 5). As the volume of data collected continues to expand, exposure scientists need to consider the most appropriate approaches for reporting information on a multitude of exposure sources, some of which might be ambiguous with regard to importance, might be sensitive, or might pose personal risks to individual participants (for example, in legal proceedings) (Deck and Kosatsky 1999; Morello-Frosch et al. 2009). In addition to returning exposure data to participants, consideration will need to be given to how to place the data into context with respect to whether there are potential health implications or a need for people to take action to reduce exposures.

At the present time there is little guidance for scientists and academic-community researchers on reporting back individual and community-level exposure data to study participants (NRC 2006). For example, few precedents exist for reporting biomonitoring data back to individuals when little information is available for interpreting health implications (Morello-Frosch et al. 2009). Morello-Frosch et al. identified distinct frameworks used by scientists for reporting back biomonitoring results. For example in the clinical approach, that is biomedically driven, reporting individual biomonitoring results is based on whether the risk relationship between exposure and health effects is understood. With CBPR, sharing of individual and community-level exposure data are encouraged between researchers and participants because it is believed that these data can have an impact beyond individual health, including in allowing communities to understand sources and potentially reduce exposures.

There appears to be movement in favor of addressing the issue of reporting-back exposure data in the recruitment and consent process of research studies. The ethical issues of reporting back exposure data will become more complex as data are increasingly integrated, reinforcing the need to address the rights of study participants before, during, and after studies are conducted. This is particularly true if exposure samples are to be stored for use after a study is completed (Morello-Frosch et al. 2009).

Engaging Affected Communities

As noted above, there are numerous challenges related to how scientists engage differentially burdened and exposed populations to address health disparities. With increasing recognition of the importance of social and ecologic stressors in human health, how we engage these populations in multiple aspects of exposure science—before, during, and after exposure studies—takes on even greater importance.

To be effective, community engagement must be multidirectional, with sources and receivers of information in constant communication with each other. Because scientists and the public may have different perceptions of risks (see, for example, Slovic et al. 2000), those responsible for communicating risks identified through exposure science need to understand stakeholders' values, concerns, and perceptions. Failure to recognize the differences and address them appropriately can cause failures in communication. In addition, the social context in which communication occurs and the extent to which sources of information are trusted play a critical role in how information is understood (Slovic et al. 2000; Miller and Solomon 2003).

Studying Susceptible and Vulnerable Populations

Ethical issues associated with exposure studies that involve susceptible and vulnerable populations—such as children, economically disadvantaged populations, the infirm, and workers—will continue to present challenges for the field of exposure science and the rest of environmental health. With the increasing reliance on chemicals in products and improvements in our knowledge about mixtures, cumulative exposures, and exposures at different life stages, the field of exposure science will be challenged to perform ethical studies that involve and will ultimately protect susceptible and vulnerable populations. There will be a greater demand for this type of research to advance the state of knowledge while balancing the needs of those populations and to perform exposure research that is beneficial and produces high-quality data that can be used to reduce and eliminate human exposures.

Protecting the Natural Environment

Protecting and maintaining the integrity and health of ecosystems is critical because human society depends on natural processes and goods provided by natural ecosystems. As discussed in Chapter 1, ecologic systems support human health and well-being. Thus, even from an anthropocentric perspective, alterations of natural ecosystems that reduce their capacity to provide ecologic services may harm current populations or future generations. This perspective is distinct from the continuing discussion within environmental ethics regarding the need to extend the traditional boundary of societal concerns, which center on human concerns, to include nonhuman elements of our world (Des Jardins 2012).

To the extent that human health and ecologic health are inextricably linked, agencies that fund exposure-assessment studies will need to consider the long-term value of funding studies that focus on ecologic and environmental consequences of exposure even when direct effects on human health may be unclear. This approach reflects growing awareness that our global ethical responsibilities extend beyond humans and include the nonhuman elements of ecosystems. In addition, environmental stewardship is essential for supporting the full array of ecosystem services on which human health and well-being depend. Although budgetary pressures may make it difficult to advocate for studies of environmental exposure, advancing exposure science requires recognition of the interconnectedness of human health and ecologic health.

GUIDING VALUES: THE RIGHT TO LEARN

In developing practical approaches to societal challenges that will increase public support for exposure science, researchers conducting exposure studies should be mindful of the guiding values that shape their research. The investigation of environmental problems is an inherently ethical undertaking entwined with competing societal and political values. A researcher's choice to investigate a particular toxicant or geographic location, for example, validates a set of societal concerns about specific risks, irrespective of the eventual outcomes of the research. In addition, the limitations of public resources for research require federal agencies and the researchers that they support to be good stewards of the resources and to choose to pursue problems that have the potential to improve public health. Studies of human exposures can create new legal and ethical obligations, both for those who have created an environmental hazard and for those who are exposed.

Human and environmental exposure-science research can seem distant to the public. Those who are familiar with exposure science have tended to view the field as driven largely by regulatory and political agendas. In addition, however, there is a need to convey to the public the rich array of information generated by human and environmental exposure science. With the development of smart phones and ubiquitous sensing, providing such information through monitoring of exposures is becoming more feasible. We elaborate on the need for transparency and engagement below, following the general theme of moving from source to dose.

1. **Sources**—We believe that people have a right to learn about sources of environmental exposures that potentially affect their health and well-being or the health of the ecosystems on which they depend. As with all research for public good, the exposure science community needs to make sure that ways are found to educate people about sources of exposures that may affect their health and well-being.

2. **Ambient and indoor concentrations**—People also have a right to know about exposure concentrations in all the microenvironments that they inhabit, and there is a reciprocal duty for researchers and others involved in the translation of scientific data to educate people about ambient concentrations that may affect their health.

3. **Personal actions and exposures**—People have a right to know how their own personal actions affect the environmental exposures that they encounter and how behavioral changes can reduce these exposures. To the extent that exposure assessment has traditionally been driven largely by regulatory purposes, there is a need to reorient the focus of exposure assessment to engage a more diverse set of stakeholders in decisions about the direction of human and environmental exposure science. Needs for translation of exposure science to educate citizens extends from informing people about how they can change their behaviors to improve personal health to informing those responsible for sources of harmful exposures how they can reduce harmful emissions and thereby contribute to improving environmental and public health.

4. **Body burdens of exposure**—With the advent of "-omics" technologies, we anticipate a wealth of information on body burdens of exposures. People who supply such information have a right to know about the body burdens of exposure that they carry, and it is incumbent on those collecting exposure data to educate people who have donated the data about their potential consequences.

Those values are consistent with another major theme in our report, namely, that exposure science needs to move beyond the assessment of ambient exposures to the characterization of dose. Emerging surveillance technologies will soon allow ubiquitous, large-scale monitoring of body burdens from multiple exposures. Although individual citizens will have different levels of interest and resources for learning about their environmental exposures, embracing the above values will help to ensure that exposure science engages a broad and diverse group of stakeholders who are interested in learning more about their personal exposures.

CONCLUSIONS

To be deserving of the public trust entails a commitment to maintaining the highest standards of integrity in exposure-science research. Scientists engaged in exposure assessment need to structure their studies in a way that is respectful of research volunteers and minimizes the potential for harm to people who make their research possible by voluntarily participating in exposure studies. When conducting participatory research, scientists should engage affected communities and populations in order to generate useful exposure information that stakeholders can use to reduce and eliminate their exposures and improve community health.

As we prepare for a time when new exposure-monitoring technologies and research paradigms expand the traditional scope of exposure-assessment studies, it is critical to anticipate ethical challenges that will present themselves. Several of the challenges stem from the collection of increasingly individualized data, which can present greater risk for research volunteers or create an expectation that results of potential importance will be returned to individual participants. The expanding scope of exposure-assessment studies may also create opportunities to provide citizen-volunteers with exposure data that they would find useful in reducing their personal exposure to environmental stressors. Scientists conducting exposure-assessment studies and the agencies that support their work should consider how to anticipate and respond to these challenges so as to maximize the global impact of human and environmental exposure science and sustain public confidence in the integrity of this evolving field of research.

- Federal agencies that support human and environmental exposure science need to develop educational programs to improve understanding of exposure-assessment research. The programs need to engage members of the general public, specialists in research oversight (such as members of IRBs), and specific communities that are differentially burdened by environmental toxicants.
- Federal agencies and professional societies should develop programs to assist exposure scientists in navigating the complex terrain of human-subjects research. Changing regulatory requirements demand that exposure scientists stay current with IRB expectations. Better communication between exposure scien-

tists and regulatory specialists can help to ensure that the expectations are transparent to those engaged in exposure-science studies.

REFERENCES

Acury, T.A., S.A. Quandt, and L. McCauley. 2001. Farmworker pesticide exposure and community-based participatory research: Rationale and practical applications. Environ. Health Perspect. 109(suppl. 3):429-434.

Baron, S.L., and S. Wilson. 2011. Occupational and environmental health equity and social justice. Pp. 69-97 in Occupational and Environmental Health: Recognizing and Preventing Disease, B.S. Levy, D.H. Wegman, S.L. Barry, and R.K. Sokas, eds. Oxford: Oxford University Press.

Brown, P., J.G. Brody, R. Morello-Frosch, J. Tovar, A.R. Zota, and R.A. Rudel. 2012. Measuring the success of community science: The northern California household exposure study. Environ Health Perspect. 120(3):326-331.

Chess, C., B.J. Hance, and P.M. Sandman. 1988. Improving Dialogue with Communities: A Short Guide for Government Risk Communication. Division of Science, Research and Technology, New Jersey Department of Environmental Protection [online]. Available: http://mss3.libraries.rutgers.edu/dlr/showfed.php?pid=rutgerslib:31712 [accessed Jan. 24, 2012].

Corburn, J. 2002. Combining community-based research and local knowledge to confront asthma and subsistence-fishing hazards in Greenpoint/Williamsburg, Brooklyn, New York. Environ. Health Perspect. 110(suppl. 2):241-248.

Deck, W., and T. Kosatsky. 1999. Communicating their individual results to participants in an environmental exposure study: Insights from clinical ethics. Environ. Res. 80(2 Pt. 2):S223-S229.

Des Jardins, J.R. 2012. Environmental Ethics, 5th Ed. Boston, MA: Wadsworth.

EPA (U.S. Environmental Protection Agency). 2003. Towards an Environmental Justice Collaborative Model: Case Studies of Six Partnerships Used to Address Environmental Justice Issues in Communities: Case Studies. EPA/100-R-03-002. Office of Policy, Economics, and Innovation, U.S. Environmental Protection Agency, Washington, DC [online]. Available: http://www.epa.gov/evaluate/pdf/ejevalcs.pdf [accessed Jan. 24, 2012].

EPA (U.S. Environmental Protection Agency). 2006. EPA's Environmental Justice Collaborative Problem-Solving Model. EPA 300-R-06-002. U.S. Environmental Protection Agency, Washington, DC [online]. Available: http://permanent.access.gpo.gov/lps87788/cps-manual-12-27-06.pdf [accessed Jan. 24, 2012].

EPA (U.S. Environmental Protection Agency). 2008. Scientific and Ethical Approaches for Observational Exposure Studies. EPA 600/R-08/062. National Exposure Research Laboratory, Office of Research and Development, U.S. Environmental Protection Agency, Research Triangle Park, NC [online]. Available: http://www.epa.gov/nerl/sots/SEAOES_doc20080707.pdf [accessed Jan. 24, 2012].

Fleming, C. 2004. When environmental justice hits the local agenda: A profile of Spartanburg and Spartanburg County, South Carolina. PM Magazine 86(5):1-10.

Graham, J.A., ed. 2010. Exposure Science Digests: Demonstrating How Exposure Science Protects Us From Chemical, Physical, and Biological Agents. Journal of Exposure Science and Environmental Epidemiology [online]. Available: http://www.nature.com/jes/pdf/JESSE_ESD_booklet.pdf [accessed Jan. 24, 2012].

Habisreutinger, P., and D.E. Gunderson. 2006. Real estate reuse opportunities within the ReGenesis project area: A case study. Int. J. Constr. Educ. Res. 2(1):53-63.

Israel, B.A., A.J. Schulz, E.A. Parker, and A.B. Becker. 1998. Review of community-based research: Assessing partnership approaches to improve public health. Annu. Rev. Public Health 19:173-202.

Kamel, F., T. Moreno, A.S. Rowland, L. Stallone, G. Ramírez-Garnica, and D.P. Sandler. 2001. Recruiting a community sample in collaboration with farmworkers. Environ. Health Perspect. 109(suppl. 3):457-459.

Lundgren, R.E., and A.H. McMakin. 2009. Risk Communication: A Handbook for Communicating Environmental, Safety and Health Risks, 4th Ed. Hoboken, NJ: John Wiley & Sons.

McCreanor, J., P. Cullinan, M.J. Nieuwenhuijsen, J. Stewart-Evans, E. Malliarou, L. Jarup, R. Harrington, M. Svartengren, I.K. Han, P. Ohman-Strickland, K.F. Chung, and J. Zhang. 2007. Respiratory effects of exposure to diesel traffic persons with asthma. N. Engl. J. Med. 357(23):2348-2358.

Miller, M., and G. Solomon. 2003. Environmental risk communication for the clinician. Pediatrics 112(suppl. 1):211-217.

Mills, N.L., H. Tornqvist, M.C. Gonzalez, E. Vink, S.D. Robinson, S. Söderberg, N.A. Boon, K.Donaldson, T. Sandström, A. Blomberg, and D.E. Newby. 2007. Ischemic and thrombotic effects of dilute diesel-exhaust inhalation in men with coronary heart disease. N. Engl. J. Med. 357(11):1075-1082.

Minkler, M., V. B. Vásquez, M. Tajik, and D. Petersen. 2006. Promoting environmental justice through community-based participatory research: The role of community and partnership capacity. Health Educ. Behav. 35(1):119-137.

Morello-Frosch, R., J.G. Brody, P. Brown, R.G. Altman, R.A. Rudel, and C. Pérez. 2009. Toxic ignorance and right-to-know in biomonitoring results communication: A survey of scientists and study participants. Environ. Health 8:6.

NRC (National Research Council). 1989. Improving Risk Communication. Washington, DC: National Academy Press.

NRC (National Research Council). 1996. Understanding Risk: Informing Decisions in a Democratic Society. Washington, DC: National Academy Press.

NRC (National Research Council). 2004. Intentional Human Dosing Studies for EPA Regulatory Purposes: Scientific and Ethical Issues. Washington, DC: National Academies Press.

NRC (National Research Council). 2006. Human Biomonitoring for Environmental Chemicals. Washington, DC: National Academies Press.

ReGenesis. 2008. ReGenesis Environmental Justice Demonstration Project: Community Revitalization through Partnerships [online]. Available: https://www.communicationsmgr.com/fs_regenesis.asp [accessed Jan. 24, 2012].

Resnik, D.B., and S. Wing. 2007. Lessons learned from the Children's Environmental Exposure Research Study. Am. J. Public Health 97(3):414-418.

Schulte, P., and M. Singal. 1996. Ethical issues in the interaction with subjects and disclosure of results. Pp. 178-198 in Ethics and Epidemiology, S. Coughlin, and T. Beauchamp, eds. New York: Oxford University Press.

Shepard, P.M., M.E. Northridge, S. Prakash, and G. Stover. 2002. Advancing environmental justice through community-based participatory research. Environ. Health Perspect. 110(suppl. 2):139-140.

Slovic, P. 2000. The Perception of Risk. London, UK: Earthscan.

Wilson, S.M., O.R. Wilson, C.D. Heaney, and J. Cooper. 2008. Community-driven environmental protection: Reducing the P.A.I.N. of the built environment in low-income African-American communities in North Carolina. Pp. 41-58 in Social Justice in Context: 2007-2008, Vol. 3. Carolyn Freeze Baynes Institute for Social Justice, East Carolina University, Greenville, NC [online]. Available: http://www.ecu.edu/che/docs/Social%20Justice%20in%20Context%200708%20Vol%203.pdf [accessed May 10, 2012].

7

Realizing the Vision

INTRODUCTION

In this report, the committee has evaluated the status of exposure science and how the field is poised to play a more critical role in addressing the important human health and ecologic challenges of the future. The committee's analyses established that exposure science is essential for protecting human and ecosystem health by informing decisions about prevention and mitigation of adverse exposures and by enabling sustainable innovations. The committee expanded the vision of exposure science to the eco-exposome. Adoption of this concept will lead to the development of a universal exposure-tracking framework that allows for the creation of an exposure narrative and the prediction of virtually all biologically-relevant human and ecologic exposures, leading to improved exposure information for making informed decisions to protect human and ecosystem health.

With better exposure information, the field has the ability to address multiple and complex scientific, societal, commercial, and policy demands. To provide the level and quality of exposure information on the scale required by those demands, the collection of relevant information—with both traditional established methods, such as pollution-monitoring networks, and emerging methods, such as those in exposure biology and in application of cellular-telephone networks—needs to be improved. In addition, it will become important to take advantage of advances in related fields of science, including biology, informatics, and microsensor technologies.

The increasingly complex global-scale interactions among the built and natural environments call for better, faster, and less-expensive exposure information. Such information is essential for managing health and environmental risks, protecting vulnerable populations, and developing innovative and sustainable solutions to prevent exposures to adverse stressors and to promote exposures to beneficial ones.

Embedded in the committee's vision is the recognition of the integrative nature of human and environmental systems. There are no boundaries between organisms (including humans) and their environment or between the internal environment of the human body and the external environment. Historically, exposure research has focused on discrete exposures—in either external or internal environments, concentrating on effects from sources on biologic systems, either human or ecologic—one stressor at a time. As a result, tools and methods evolved, and resources were channeled to address specific measures.

To fulfill its vision, the committee has identified the following overarching research needs in exposure science:

- Characterizing exposures quickly and cost-effectively at multiple levels of integration—including time, space, and biologic scales—and for multiple and cumulative stressors.
- Scaling up methods and techniques to detect exposure in large human and ecologic populations of concern.

The broader availability and ease of use of technologies, including sensor, analytic, bioinformatic, and computational technologies, have given rise to a substantial profusion of data and an overall democratization of the collection and availability of exposure information. The Centers for Disease Control and Prevention (CDC) National Human and Nutrition Examination Survey (NHANES) provides one of the most revealing snapshots of human exposures to over 200 environmental chemicals through the use of biomonitoring (CDC 2011). The collaboration between CDC and national and international organizations quickly expanded the breadth and depth of data available throughout populations and subpopulations (NRC 2006). That rapid progress was predicated on the availability of better analytic methods and a national commitment to generate such "baseline" data.

With the availability of the emerging measurement and informatics technologies, the committee sees both the demand and the opportunity for conducting strategic data-gathering efforts to answer a multitude of environmental-exposure questions. Such efforts could involve deploying large numbers of environmental sensors and networking technologies and collecting biomonitoring samples in statistically representative populations. The resulting data could be integrated with informatics capabilities for collection, storing, and analyzing the information gathered and used to test environmental-health–related hypotheses or to develop exposure-reduction strategies.

The committee recognizes that realizing its vision requires an iterative approach that will initially develop and implement innovative tools to meet the urgent demands for exposure information today while establishing the infrastructure, including educational opportunities and study sections devoted to research, needed to transform the science fully over the next 20 years. This chapter describes a pragmatic approach to realizing the vision for exposure science in the 21st century whereby resources are deployed to generate and analyze the maximum amount of exposure information and to develop effective and relevant applications of such information. One important objective should be to describe, reconstruct, and forecast real-world exposures more accurately and more efficiently. To be effective, exposure science needs to adopt a systems-based approach that, to the extent possible, considers exposures from source to dose and from dose to source and considers multiple levels of integration, including time, space, and biologic scales in connection with multiple stressors in human and ecosystem populations.

THE EXPOSURE DATA LANDSCAPE[1]

In the near term, exposure science needs to develop strategies to expand exposure information rapidly to improve understanding of where, when, and how exposures occur and their health significance. Data generated and collected would be used to evaluate and improve models of exposure for use in generating hypotheses and developing policies. New exposure infrastructure (for example, sensor networks, environmental monitoring, activity tracking, and data storage and distribution systems) will help to refine or replace existing measurement and monitoring strategies. This process will help to identify the largest knowledge gaps and reveal where gathering of more exposure information would contribute the most to reducing uncertainty.

In the field of environmental health, substantial investment and progress have been made in recent years to collect and improve access to genomic, toxicology, and health data (for example, Davis et al. 2011; CTD 2012) and to provide information on chemical toxicity and inform and guide research. However, those data have historically lacked the extensive and reliable exposure information required for examining environmental contributions to diseases and assessing health risks. The Environmental Protection Agency (EPA) ExpoCast program, initiated to address that research gap, is intended to advance the characterization of exposures to support toxicity testing (Cohen Hubal et al. 2010a) and in the long term to link exposures to health outcomes. There is still a growing demand to collect more exposure data to populate emerging exposure databases (for example, Gangwal 2011) and to facilitate linkages with toxicity and environmental-fate data and with manufacturers' production and use data.

An Exposure Infostructure

Exposure data are often scattered among such widely dispersed sources that it is difficult to relate them (Egeghy et al. 2012). Several initiatives in the United States and abroad aim at developing tools to integrate those data sources

---

[1]"Data landscape" is a term used in informatic and computational analyses. The term implies stepping back and looking at the data available, identifying data rich and data poor areas, and seeing what the data "landscape" looks like.

(for example, Mattingly et al. 2012), but more efforts are needed. The tools will help in the systematic mining of databases and scientific literature that contain millions of observations and are intended to bridge the gap between exposure science and other environmental-health disciplines, including toxicology, epidemiology, disease surveillance, and epigenetics.

Efforts to develop and enhance exposure information can contribute to the development of an exposure infostructure and data-sharing approaches that in turn can influence the design of exposure and environmental-health studies. In such fields as genomics, computational toxicology, environmental toxicology, and cancer research, creation of infostructures has resulted in groundbreaking and transformative innovations in research methods and approaches. A strategic approach to populating the exposure knowledge base effectively will motivate research that informs our understanding of exposures at biologic scales, time durations, and locations. Consideration of the uncertainty and variability of measurements and models is critical because data will be gathered from disparate sources and will influence our understanding of health and ecosystem effects.

Developing and Empowering Computational Approaches

Environmental fate and transport models are used to estimate and predict environmental concentrations. However, the models are hampered by the absence of data that can be used to evaluate some of the model parameters and ultimately estimate the relevance and robustness of their predictions (MacLeod et al. 2010). Data in the exposure infostructure can be used to cross-validate the models against one another and bridge knowledge gaps. The models can then be developed further to predict exposures to a large number of chemicals or other stressors individually or in combination and among microenvironments. For example, Aylward and Hays (2011) used data from the NHANES biomonitoring program and pharmacokinetic studies, integrated with in vitro toxicity data from specific chemical case studies, to examine the physiologic relevance of tested in vitro concentrations and thereby helped to inform dosimetry in evaluating ToxCast data. Efforts to determine how exposure models could be adapted to advance high-throughput chemical priority-setting and risk assessment have been hindered in part by absence of or lack of access to high-quality exposure data (Cohen Hubal et al. 2010a; Egeghy et al. 2012).

Integrating Surveillance Systems

Responding to the demand for exposure data requires creative and parsimonious approaches for data generation and collection. The opportunities provided by existing and emerging surveillance networks and technologies to gather direct exposure information or information on relevant surrogates can improve our assessment of exposures, for example, by combining food samples and soil samples and linking ecologic surveys with food-web analysis to evaluate exposure models for complex food-pathway exposures. In the near term, an important research goal will be to identify the diversity of existing surveillance systems to improve our knowledge of data on environmental contaminants. Specifically, researchers will need to address the following questions: How can these surveillance systems be used to provide baseline information on exposure? How can resources be marshaled to obtain such data? What efforts can be made to develop surveillance systems, in particular ones that will integrate ecologic surveillance data with human health data. One large-scale example is the CDC National Environmental Public Health Tracking Network (National EPHT Network), which collects information relevant for assessing environmental exposures and associated health outcomes and develops the infrastructure needed for analyzing and integrating such information for public-health protection. That effort faces many challenges—the network includes 23 states—but the National EPHT Network constitutes the largest effort to track environmental exposures that are likely to contribute to disease.

When hot spots or places of highest potential impact to vulnerable and susceptible populations are identified, targeted studies need to be designed for more detailed measurements. That will expand applications of exposure science to specific studies that can lead to exposure reduction or source mitigation. It can also include the development of new tools.

In essence, advances in technologies and bioinformatics provide a plethora of opportunities to link existing surveillance systems and data infrastructures and to enrich them with targeted exposure-measurement studies that will promote the development of an exposure infostructure that increases our understanding of health and ecologic impacts of environmental exposures.

A Predictive Exposure Network

The combination of surveillance programs and targeted exposure-measurement programs is integral to the strategy for building a predictive exposure network that can address environmental-health questions. Information from the network could be used to develop exposure metrics that will provide the information needed for evaluating the overall health and resilience of humans and ecosystems, identifying vulnerable populations, assessing the impact of cumulative exposures, and addressing exposure disparities. It could also be used to assess environmental improvements and to provide early warnings of emerging problems. More data on exposures will allow us to forecast, prevent, and mitigate the impacts of such major societal challenges as climate change, security threats, and urbanization.

Given the explosion of technologies and knowledge systems, this incremental, iterative, and adaptive approach to developing a network is feasible even in a resource-constrained environment. It will require modest resources and a commitment from the community of exposure scientists.

IMMEDIATE CHALLENGES: CHEMICAL EVALUATION AND RISK ASSESSMENT

A major demand for exposure information comes from efforts to modernize chemical-management policies in the United States and abroad, including the European Commission's Registration, Evaluation, and Authorization of Chemicals (REACH) regulation, the Green Chemistry Initiative in California, the sustainability program in EPA, and efforts to revise the Toxic Substances Control Act. Those efforts have highlighted the need for tools for assessing and measuring exposures to a large number of chemicals currently on the market and others that are emerging and likely to become ubiquitous, such as nanomaterials. In addition, there is a need for improved understanding of multiple exposures and tools for assessing biologically relevant exposures, particularly during critical life stages. The confluence of interests with recent advances in biology, toxicology, and computational tools provides opportunities to advance exposure assessment.

Setting Priorities among Exposures

There is a need to characterize potential risk to human and ecosystem health that arises from the manufacture and use of tens of thousands of chemicals (Cohen Hubal 2010b). EPA's ToxCast program is applying new technologies to screen and set priorities among chemicals for toxicity. EPA has developed methods for using high-throughput screening and toxicogenomic technologies to predict potential toxicity and to set priorities for the use of testing resources (Cohen Hubal 2008). In a parallel effort, ExpoCast is aiming to develop the required exposure-science data and tools for addressing immediate needs for rapid characterization of exposure potential for priority-setting and chemical-risk management. Through ExpoCast, EPA's Office of Research and Development aims to develop novel approaches and metrics for chemical screening and evaluation based on biologically relevant human exposures (Little et al. 2011).

EPA's National Center for Computational Toxicology has proposed a Toxicological Priority Index (ToxPi) designed to integrate multiple domains of knowledge to inform chemical priority-setting (EPA 2011). The ToxPi framework is flexible, can incorporate new data from diverse sources, and provides an opportunity to enrich priority-setting related to potential hazards with exposure information (Reif et al. 2010; Little et al. 2011).

Combining exposure priority-setting information with hazard information, such as that derived from ToxCast, will help in establishing priorities among chemicals for evaluation on the basis of their potential for harming human health. With that information, it will be possible to develop exposure assessments that can identify the appropriate type of information and the level of detail needed to address the risk-assessment and risk-management questions at hand.

Assessing and Quantifying Multiple Exposures

To assess the outcomes of multiple exposures (that is, both exogenous and endogenous stressors), there is a need to understand the joint behavior of these stressors, the interactions among them, and their contributions to health outcomes. This includes research to address interactions among chemical, physical, and biologic stressors, along with social stressors. Understanding the sources of these stressors can allow for intervention to prevent exposures or to mitigate their effects.

Although it is possible to test the toxicity of mixtures of chemicals (and perhaps other stressors), the tests tend to be based on ad hoc combinations typically of two chemicals and are often not very representative of real-world exposures. In a recent analysis, researchers in EPA investigated methods from the field of community ecology originally developed to study avian species co-occurrence patterns and adapted them to examine chemical co-occurrence (Tornero-Velez et al. 2012). Their findings showed that chemical co-occurrence was not random but was highly structured and usually resulted in specific combinations that were predictable with models. Novel application of tools and approaches from a variety of research disciplines can be used to address the complexity of mixtures, advance our understanding of exposures to them, and promote the design of relevant experiments and models to assess their health risks.

Characterizing or Quantifying Biologically Relevant Exposures

Systems approaches to understanding human biology together with knowledge of systems-level perturbations caused by human–environment interactions are critical for understanding biologically relevant exposures (Farland 2010). Understanding how early perturbations of biologic pathways can lead to disease requires information gathered over a lifetime. In addition, applying such concepts as the exposome effectively demands exposure information that is predictive of disease. The connection between exposure information for understanding early perturbations of biologic pathways and for predicting disease carries enormous promise for better ways of linking exposure and disease and ultimately for informing design of relevant studies. The development of advanced technologies to measure key exposure metrics that include biomarkers for assessing internal exposures and sensors to measure personal exposure needs to be supported to achieve a better understanding of exposure–response relationships (Cohen Hubal 2009). Integrated application of the technologies in specific situations will help to elucidate the exposures that are relevant to biologic effects of environmental hazards. Such applications will allow us to assess the effects of aggregate and cumulative exposures on health.

Such efforts are currently funded under the National Institute of Environmental Health Sciences (NIEHS) Exposure Biology Program[2] (NIEHS 2009), and related efforts are funded by EPA and other federal agencies. These novel biomarker technologies are still in their infancy and require resources for developing them, scaling them, and validating them for population studies. An incremental and iterative approach to creating opportunities for transdisciplinary research, cross-study sharing, and validation of tools and data will help to advance their progress.

IMPLEMENTING THE VISION

Exposure information is needed to advance environmental-health research. Because of the relative scarcity of exposure data and the high cost of collecting them, environmental-health analyses and decisions have often been based on narrowly limited or low-quality data. However, as discussed earlier (see Chapter 1), the absence of such data has had unintended consequences (for example, Graham 2011). The demand for exposure information, coupled with the development of tools and approaches for collecting and analyzing such data, has created an opportunity to transform exposure science to advance human and ecosystem health.

The transformation will require an investment of resources and a substantial shift in how exposure science research is deployed and implemented.

RESEARCH NEEDS

To implement its vision, the committee identified research needs that call for new methods and approaches, validation of methods and their enhancement for application on different scales and in broader circumstances, and improved linkages to research in other sectors of the environmental-health sciences. The research needs are organized into several broad categories: providing effective responses to immediate or short-term threats to health and the environment; supporting research on health and ecologic effects to understand past, current, and emerging outcomes; and addressing demands for exposure information from communities, government, and industry. The research needs are organized by priority within each category on the basis of the time that will be required for their development and implementation: *short term* denotes less than 5 years, *intermediate term* 5–10 years, and *long term* 10–20 years.

---

[2]The exposure biology program is investing in new technologies to assess how environmental exposures, including diet, physical activity, stress, and drug use, contribute to human disease. This includes sensors for chemicals in the environment, new ways to characterize dietary intake, levels of physical activity, responses to psychosocial stress, and measures of the biologic response to these factors at the physiologic and molecular levels (NIEHS 2009).

Providing effective responses to immediate or short-term public-health or ecologic risks requires research on observational methods, data management, and models:

Short term

- Identify, improve, and test instruments that can provide real-time tracking of biologic, chemical, and physical stressors to monitor community and occupational exposures to multiple stressors during natural, accidental, or terrorist events or during combat and acts of war.
- Explore, evaluate, and promote the types of targeted population-based exposure studies that can provide information needed to infer the time course of internal and external exposures to high-priority chemicals.

Intermediate term

- Develop informatics technologies (software and hardware) that can transform exposure and environmental databases that address different levels of integration (time scales, geographic scales, and population types) into formats that can be easily and routinely linked with population-wide outcome databases (for humans and ecosystems) and linked to source-to-dose modeling platforms to facilitate rapid discovery of new hazards and to enhance preparedness and timely response.
- Identify, test, and deploy extant remote sensing, high-volume personal monitoring techniques, and source-to-dose model-integration tools that can quantify multiple routes of exposure (inhalation, ingestion, and dermal uptake) and obtain results that can, for example, be integrated with emerging methods (such as –omics technologies) for tracking internal exposures.

Long term

- Enhance tracking of human exposures to pathogens on the basis of a holistic ecosystem perspective from source through receptor.

Supporting research on health and ecologic effects that addresses past, current, and emerging outcomes:

Short term

- Coordinate research with human-health and ecologic-health scientists to identify, collect, and evaluate data that capture internal and external markers of exposure in a format that improves the analysis and modeling of exposure–response relationships and links to high throughput toxicity testing.

- Explore options for using data obtained on individuals and populations through market-based and product-use research to improve exposure information used in epidemiologic studies and in risk assessments.

Intermediate term

- Develop methods for addressing data and model uncertainty and evaluate model performance to achieve parsimony in describing and predicting the complex pathways that link sources and stressors to outcomes.
- Improve integration of information on human behavior and activities for predicting, mitigating, and preventing adverse exposures.

Long term

- Adapt hybrid designs for field studies to combine individual-level and group-level measurements for single and multiple routes of exposure to provide exposure data of greater resolution in space and time.

Addressing demands for exposure information among communities, governments, and industries with research that is focused, solution-based, and responsive to a broad array of audiences:

Short term

- Develop methods to test consumer products and chemicals in premarketing controlled studies to identify stressors that have a high potential for exposure (intake fraction) combined with a potential for toxicity to humans or ecologic receptors.
- Develop and evaluate cost-effective, standardized, non-targeted, and ubiquitous methods for obtaining exposure information to assess trends, disparities among populations (human and ecologic), geographic hot spots, cumulative exposures, and predictors of vulnerability.

Intermediate term

- Apply adaptive environmental-management approaches to understand the linkages between adverse exposures in humans and ecosystems better.
- Implement strategies to engage communities, particularly vulnerable or hot-spot communities, in a collaborative process to identify, evaluate, and mitigate exposures.

Long term

- Expand research in ways to use exposure science to more effectively regulate environmental risks in natural and human systems, including the built environment.

TRANSAGENCY COORDINATION

Exposure science is relevant to the work and mission of many federal agencies. A transagency collaboration for exposure science in the 21st century would accelerate progress in and transform the field.

Tox21 is a collaboration among EPA, the National Institutes of Health (NIH), and recently the Food and Drug Administration (Collins et al. 2008; Schmidt 2009). The collaboration has been extended to include research partners in Europe and Asia. It resulted from the National Research Council recommendations (NRC 2007) that called for a long-range vision for transforming toxicology to meet the demands of the 21st century—not unlike the vision offered in the present report. The primary objective of the Tox21 collaboration was to leverage resources and expertise. It included sharing databases and analytic tools, cataloging critical toxicologic data for key target organs, sponsoring workshops to broaden scientific input into strategy and direction, engaging the international community, and promoting scientific training and outreach. The budget for Tox21 was developed gradually on the basis of the success of the initial research, and the momentum created by this effort influenced research planning and budgetary directions for other organizations, including industry, nongovernmental organizations, and other federal agencies, to bring resources and expand on this collaboration.

The present committee considers that the model used in establishing Tox21 should be extended to exposure science. This would create Exposure21. In addition to the engagement of those stakeholders involved in Tox21, engagement of other federal agencies—such as the US Geological Survey, CDC, the National Oceanic and Atmospheric Administration, the National Science Foundation (NSF), and the National Aeronautics and Space Administration—would promote access to and sharing of data and resources on a broader scale. Including them would provide access to resources for transformative technology innovations, for example, in nanosensors.

ENABLING RESOURCES

The research needs discussed in the report extend to the activities and the mandates of individual agencies, including EPA and NIEHS. The programmatic activities of these agencies will be improved by embracing new basic and applied exposure-science research. Over time, the results of this research will provide opportunities to demonstrate its value to aligned agencies and will lead to the formation of new partnerships in exposure science. Such collaborative transformations will improve the ability of decision-makers to use the results in risk-management and in risk-prevention programs. Thus, the committee recommends that intramural and extramural programs at the EPA, NIEHS, Department of Defense, and other agencies that advance exposure-science research be supported, as the value of the research and the need for exposure information become more apparent.

Much of the human-based research in environmental-health sciences is funded by NIH. However, none of the existing study sections that review grant applications has substantial expertise in exposure science and most study sections are organized around disease processes. As part of stakeholder engagement in its 2011 strategic review process, NIEHS identified exposure science as a subject of high research priority (NIEHS 2011a). In light of this new emphasis and the role that an understanding of environmental exposures can play in disease prevention, a rethinking of how NIH study sections are organized that incorporates a greater focus on exposure science would allow a core group of experts to foster the objectives of exposure-science research. In addition, research collaborations between agencies could leverage resources for expanded exposure-science research; for example, collaborations among EPA, NIEHS, and NSF could support integrative research between ecosystem and human-health approaches in exposure science. However, many other agencies engaged in exposure science research could be included in the collaborations.

An additional concern is the need to educate the next generation of exposure scientists or to provide opportunities for members of other fields to cross-train in the techniques and models used to analyze and collect exposure data. The effective implementation of the committee's vision will depend on development and cultivation of scientists, engineers, and technical experts. For years, academic institutions have mostly trained exposure scientists on the periphery of other programs, such as industrial hygiene and epidemiology. To implement the vision, a new crop of transdisciplinary scientists will need to be trained with integrated expertise in many fields of science, technology, and environmental health, with a focus on problem formulation and solution-based approaches. Exposure scientists will need the skills to collaborate closely with other fields of expertise, including engineers, epidemiologists, molecular and systems biologists, clinicians, statisticians, and social scientists. To achieve that, the committee considers that the following is needed

- An increase in the number of academic predoctoral and postdoctoral training programs in exposure science throughout the United States supported by training grants. NIEHS currently funds one training grant in exposure science; additional training grants are needed (NIEHS 2011b).

- Short-term training and certification programs in exposure science for midcareer scientists in related fields.
- Development, by federal agencies that support human and environmental exposure science, of educational programs to improve public understanding of exposure-assessment research. The programs would need to engage members of the general public, specialists in research oversight, and specific communities that are disproportionately burdened by known environmental stressors.

Participatory and Community-Based Research Programs

Responsiveness as articulated in the committee's vision involves engaging broader audiences, including the public, in ways that contribute to problem formulation, monitoring and collection of data, ensuring access to data, development of decision-making tools, and ultimately empowerment of communities to participate in reducing and preventing exposures and addressing environmental disparities. The development of more user-friendly and less expensive monitoring equipment can allow trained people in communities to collect and upload their own data in partnership with researchers and thereby improve the value of the data collected and make more data available for purposes of priority-setting and to inform policy. One approach would be to develop pilot programs in which the communities in two large American cities are engaged in implementing a system of embedded and participatory sensors based on ubiquitous and pervasive technologies. The pilot programs would evaluate the feasibility of such systems to develop community-based exposure data that are reliable and the ability of communities to use the data to understand and improve their environmental health. Potential issues of privacy would need to be considered. Examples of such efforts are described in Chapter 5.

CONCLUSIONS

The field of exposure science in environmental health began to grow from its roots in occupational health during the early 1990s with the publication of the first National Research Council report on exposure, the formation of a professional society and a journal devoted primarily to exposure science, and the publication of a number of manuscripts on the field's pivotal role in the environmental-health sciences. The committee has illustrated numerous successes in addressing environmental-health problems and shown the evolution of new tools to address current problems. The critical nature of the field illustrated in Figure 1-2 shows the centrality of efforts to mitigate the effects of sources and to intervene or prevent disease. However, as shown in the committee's vision (Chapter 2), there is still much work to do to mitigate the potential and actual effects of stressors on humans and ecosystems (for example, nanoparticles, energy systems and sources, and consumer products). Tools are evolving to determine internal and external exposures, to examine the behaviors that lead to contact, and to characterize stressors before adverse effects occur. With focus, good science, and sustained support for research and development, exposure science will have a bright future.

REFERENCES

Aylward, L.L., and S.M. Hays. 2011. Consideration of dosimetry in evaluation of ToxCast™ data. J. Appl. Toxicol. 31(8):741-751.
CDC (Centers for Disease Control and Prevention). 2011. National Report on Human Exposure to Environmental Chemicals. Centers for Disease Control and Prevention [online]. Available: http://www.cdc.gov/exposurereport/ [accessed Dec. 7, 2011].
Cohen Hubal, E. 2008. Exposure consideration for chemical prioritization and toxicity testing. Epidemiology 19(suppl. 6):S100 [Abstract ISEE-68].
Cohen Hubal, E.A. 2009. Biologically relevant exposure science for 21st century toxicity testing. Toxicol Sci. 111(2):226-232.
Cohen Hubal, E.A., A. Richard, L. Aylward, S. Edwards, J. Gallagher, M.R. Goldsmith, S. Isukapalli, R. Tornero-Velez, E. Weber, and R. Kavlock. 2010a. Advancing exposure characterization for chemical evaluation and risk assessment. J. Toxicol. Environ. Health B Crit. Rev. 13(2-4):299-313.
Cohen Hubal, E.A., A.M. Richard, I. Shah, J. Gallagher, R. Kavlock, J. Blancato, and S.W. Edwards. 2010b. Exposure science and the U.S. EPA National Center for Computational Toxicology. J. Expo. Sci. Environ. Epidemiol. 20(3):231-236.
Collins, F.S., G.M. Gray, and J.R. Bucher. 2008. Transforming environmental health protection. Science 319(5865):906-907.
CTD (Comparative Toxicogenomics Database). 2012. CTD [online]. Available: http://ctdbase.org/ [accessed Jan. 11, 2012].
Davis, A.P., B.L. King, S. Mockus, C.G. Murphy, C. Saraceni-Richards, M. Rosenstein, T. Wiegers, and C.J. Mattingly. 2011. The Comparative Toxicogenomics Database: Update 2011. Nucleic Acids Res. 39(Database issue):D1067-D1072.
Egeghy, P.P. R. Judson, S. Gangwal, S. Mosher, D. Smith, J. Vail, and E.A. Cohen Hubal. 2012. The exposure data landscape for manufactured chemicals. Sci. Total Environ. 414(1):159-166.
EPA (U.S. Environmental Protection Agency). 2011. Computational Toxicology Research Program: ToxPi. U.S. Environmental Protection Agency [online]. Available: http://www.epa.gov/ncct/ToxPi/ [accessed Dec. 27, 2011].
Farland, W.H. 2010. The promise of exposure science and assessment. J. Expo. Sci. Environ. Epidemiol. 20(3):225.
Gangwal, S. 2011. ExpoCastDB: A Publicly Accessible Database for Observational Exposure Data. Presented at Computational Toxicology Community of Practice, September 22, 2011 [online]. Available: http://www.epa.gov/ncct/download_files/chemical_prioritization/ExpoCastDB_CommPractice_09-22-2011-Share.pdf [accessed Dec. 7, 2011].
Graham, J.A., ed. 2011. Exposure Science Digests: Demonstrating How Exposure Science Protects Us From Chemical, Physical, and Biological Agents. Journal of Exposure Science and Environmental Epidemiology [online]. Available: http://www.nature.com/jes/pdf/JESSE_ESD_booklet.pdf [accessed Dec. 29, 2011].

Little, J.C., C. J. Weschler, W.W Nazaroff, Z. Liu, and E. A. Cohen Hubal. 2011. Semi-Volatile Organic Compounds in the Indoor Environment – Characterizing and Prioritizing Exposure and Risk. Presentation at SVOCs in the Indoor Environment: Mechanistic Insights to Support Sustainable Product Design, Safe Use and Improved Public Health, January 5-7, 2011, Research Triangle Park, NC [online]. Available: http://www.indair.org/index_files/Page277.htm [accessed Jan. 11, 2012].

MacLeod, M., M. Scheringer, T.E. McKone, and K. Hungerbühler. 2010. The state of multimedia mass-balance modeling in environmental science and decision-making. Environ. Sci. Technol. 44(22):8360-8364.

Mattingly, C.J., T.E. McKone, M.A. Callahan, J.A. Blake, and E.A. Cohen-Hubal. 2012. Providing the missing link: The exposure science ontology ExO. Environ. Sci. Technol. 46(6):3046-3053.

NIEHS (National Institute of Environmental Health Sciences). 2009. Exposure Biology [online]. Available: http://www.niehs.nih.gov/research/supported/programs/exposure/index.cfm [accessed Dec. 27, 2011].

NIEHS (National Institute for Environmental Health Sciences). 2011a. New Strategic Plan Process for 2012-2017 Strategic Planning [online]. Available: http://www.niehs.nih.gov/about/od/strategicplan/index.cfm [accessed Dec. 7, 2011].

NIEHS (National Institute of Environmental Health Sciences). 2011b. Training Grant for the Joint Graduate Program in Exposure Science (No. 13). NIEHS Grant No. T32ES019854 [online]. Available: http://tools.niehs.nih.gov/portfolio/index.cfm/portfolio/searchResults/grant_number/t32 [accessed Apr. 11, 2012].

NRC (National Research Council). 2006. Human Biomonitoring for Environmental Chemicals. Washington, DC: National Academies Press.

NRC (National Research Council). 2007. Toxicity Testing in the 21st Century: A Vision and a Strategy. Washington, DC: National Academies Press.

Reif, D.M., M.T. Martin, S.W. Tan, K.A. Houck, R.S. Judson, A.M. Richard, T.B. Knudsen, D.J. Dix, and R.J. Kavlock. 2010. Endocrine profiling and prioritization of environmental chemicals using ToxCast data. Environ. Health Perspect. 118(12):1714-1720.

Schmidt, C.W. 2009. Tox 21: New dimensions of toxicity testing. Environ. Health Perspect. 117(8):A348-A353.

Tornero-Velez, R., P.P. Egeghy, and E.A. Cohen Hubal. 2012. Biogeographical analysis of chemical co-occurrence data to identify priorities for mixtures research. Risk Anal. 32(2):224-236.

Appendix A

Biographic Information on the Committee on Human and Environmental Exposure Science in the 21st Century

Kirk R. Smith (*Chair*) is professor of global environmental health and director of the Global Health and Environment Program at the University of California, Berkeley School of Public Health. Previously, he was founder and head of the Energy Program of the East–West Center in Honolulu. Dr. Smith's research interests include environmental and health issues in developing countries, particularly those related to health-damaging and climate-changing air pollution from household energy use. His research also includes field measurement and health-effects studies in India, China, Nepal, Mexico, and Guatemala and development and application of tools for international policy assessments. He develops and deploys small, smart, and inexpensive microchip-based monitors for use in those settings. Dr. Smith serves on several national and international scientific advisory committees, including the Global Energy Assessment, the National Research Council's Board on Atmospheric Sciences and Climate, the executive committee for the World Health Organization Air Quality Guidelines, and the International Comparative Risk Assessment. He participated in the Intergovernmental Panel on Climate Change third and fourth assessments and thus shared the 1997 Nobel Peace Prize. Dr. Smith was elected a member of the National Academy of Sciences in 1997. He received the Heinz Prize in Environment in 2009. Dr. Smith received a PhD in biomedical and environmental health science from the University of California, Berkeley.

Paul J. Lioy (*Vice Chair*) is a professor in and the vice chair of the Department of Environmental and Occupational Medicine at the University of Medicine and Dentistry of New Jersey (UMDNJ)–Robert Wood Johnson Medical School (RWJMS). He is also the deputy director of government relations and director of exposure science at the Environmental and Occupational Health Sciences Institute of UMDNJ-RWJMS and Rutgers, the State University of New Jersey. Dr. Lioy is a member of the US Environmental Protection Agency Science Advisory Board and has served on the Board on Toxicology and Environmental Studies of the National Research Council. He is a fellow of the Collegium Ramazzini and was a member of the International Joint Commission Air Quality Board for the United States and Canada. He is a former president of the International Society of Exposure Science and was the 1998 recipient of the Wesolowski Award for Human Exposure Research. He was also the 2003 recipient of the Air and Waste Management Association Frank Chambers Award for Lifetime Research and Applications in Air Pollution and, among his other awards, was the 2008 recipient of the Rutgers Graduate School's Distinguished Alumnus Award in Mathematics, Engineering and Physical Sciences. Dr. Lioy's research interests include human exposure to environmental and occupational pollution, multimedia exposure issues for metals and pesticides, research on air-pollution exposure and dose relationships, and participation in the study of exposure and effects of pollution on human health in urban and nonurban areas and controlled environments. He is an author of 250 peer-reviewed papers and is an Information Sciences Institute Most Highly Cited Scientist in environment and ecology. Dr. Lioy has been a member of numerous editorial boards, including his current positions as associate editor of *Environmental Health Perspectives* and the *Journal of Exposure Science and Environmental Epidemiology*. He has served as a member of numerous National Research Council committees and chaired the 1987–1991 Committee on Air Pollution Exposure Assessment. Dr. Lioy received a PhD in environmental sciences from Rutgers, the State University of New Jersey.

Richard T. Di Giulio serves as director of Duke University's Integrated Toxicology Program and the Superfund Basic Research Center. His research concerns basic studies of mechanisms of contaminant metabolism, adaptation, and toxicity and the development of mechanistically based indexes of exposure and toxicity that can be used in biomonitoring. The long-term goals of his research are to bridge the gap between mechanistic toxicologic research and the development of useful tools for environmental assessment and to elucidate linkages between human and ecosystem health. The bulk of Dr. Di Giulio's work uses a comparative approach with aquatic animals, particularly fishes, as models. Of particular concern are mechanisms of oxidative metabolism of aromatic hydrocarbons, mechanisms of free-radical production and antioxidant defense, mechanisms of chemical carcinogenesis, and developmental perturbations and adaptations to contaminated environments by fishes. He received a PhD from the Virginia Polytechnic Institute and State University.

J. Paul Gilman is senior vice president and chief sustainability officer for Covanta Energy. Previously, he served as director of the Oak Ridge Center for Advanced Studies and as assistant administrator for research and development in the US Environmental Protection Agency. He also worked in the Office of Management and Budget, where he had oversight responsibilities for the US Department of Energy (DOE) and all science agencies. In DOE, he advised the secretary of energy on scientific and technical matters. From 1993 to 1998, Dr. Gilman was the executive director of the Commission on Life Sciences and the Board on Agriculture and Natural Resources of the National Research Council. He has served on the National Research Council Board on Environmental Studies and Toxicology and on several committees and in other activities of the National Research Council. Dr. Gilman received his PhD in ecology and evolutionary biology from the Johns Hopkins University.

Michael Jerrett is a professor and chair of environmental health sciences at the University of California, Berkeley School of Public Health. Since 2001, he has participated in the American Cancer Society Particle Epidemiology Project. His research interests include the spatial analysis of disease–exposure associations using geographic information science, geographic exposure modeling, and land-use characterization. Dr. Jerrett also studies environmental accounting, focusing on the determinants and evaluation of environmental costs and benefits. He has designed and analyzed local, provincial, state, and national health and environment databases in North America, Europe, and Asia. His work opened important field research connecting social determinants of health, air-pollution health effects, and spatial analysis. The spatial analysis demonstrated that the health effects of air pollution are reduced but not eliminated by ecologic (population-based) confounding and are often modified by individual and neighborhood social characteristics. Dr. Jerrett received his PhD in geography from the University of Toronto (Canada).

Petros Koutrakis is a professor of environmental sciences and director of the Environmental Chemistry Laboratory of Harvard University. He is also the director of the US Environmental Protection Agency–Harvard University Ambient Particle Center. Dr. Koutrakis is the past technical editor-in-chief of the *Journal of the Air & Waste Management Association*. His research interests include human exposure assessment, ambient and indoor air pollution, environmental analytic chemistry, and environmental management. He has more than 170 peer-reviewed publications and seven patents and has conducted a number of comprehensive air-pollution studies in the United States, Canada, Spain, Chile, and Greece. He is a member of several national and international committees and has served as a member of the National Research Council Committee on Research Priorities for Airborne Particulate Matter and the Committee for Review of the Army's Enhanced Particulate Matter Surveillance Project Report. Dr. Koutrakis received a PhD in environmental chemistry from the University of Paris.

Appendix A

Thomas E. McKone is a senior staff scientist and deputy department head at the Lawrence Berkeley National Laboratory and an adjunct professor and researcher at the University of California, Berkeley School of Public Health. Dr. McKone was appointed by California Governor Arnold Schwarzenegger to the California Scientific Guidance Panel. He is a Fellow of the Society for Risk Analysis, former president of the International Society of Exposure Analysis, and a member of the Organizing Committee for the International Life-Cycle Initiative, which is a joint effort of the UN Environment Programme and the Society for Environmental Toxicology and Chemistry. Dr. McKone's research interests include the use of multimedia compartment models in health-risk assessments, chemical transport and transformation in the environment, and measuring and modeling the biophysics of contaminant transport from the environment into the microenvironments with which humans have contact and across the human–environment exchange boundaries—skin, lungs, and gut. One of Dr. McKone's most recognized achievements was his development of the CalTOX risk-assessment framework for the California Department of Toxic Substances Control. He has been a member of several National Research Council committees, including the Committee on Environmental Decision Making: Principles and Criteria for Models, the Committee on Improving Risk Analysis Approaches Used by the U.S. EPA, and the Committee on Human Health Reassessment of TCDD and Related Compounds. He received his PhD in engineering from the University of California, Los Angeles.

James T. Oris is a professor in the Department of Zoology and is the associate provost for research and dean of the graduate school at Miami University in Oxford, Ohio. Dr. Oris's research centers on the ecologic toxicology of chemicals in aquatic systems. He has focused on sediment toxicity, photoinduced toxicity, long-term reproductive toxicity, routes of uptake, and environmental factors that may alter fate and effects. Those studies have ranged from the use of molecular biomarkers to landscape-scale ecologic assessments. Dr. Oris is also interested in standard toxicity-test development and methods, including the statistical modeling and analysis of toxicity dose–response relationships. Dr. Oris served as the president (2004–2005) of SETAC North America, a unit of the Society of Environmental Toxicology and Chemistry. He received a PhD in environmental toxicology and fisheries and wildlife from Michigan State University.

Amanda D. Rodewald is professor of wildlife ecology in the School of Environment and Natural Resources of Ohio State University. Dr. Rodewald's research program examines the mechanisms guiding landscape-scale responses of animal communities to anthropogenic disturbances on multiple spatial scales and across multiple levels of biologic organization. Her research touches on a variety of subdisciplines, including conservation biology, landscape ecology, population demography, community ecology, behavioral ecology, and ecologic restoration. She serves on the editorial board of *Studies in Avian Biology* and is a member of the Environmental Protection Agency Science Advisory Board. She received her PhD in ecology from Pennsylvania State University.

Susan L. Santos is an assistant professor in the Department of Health Education and Behavioral Sciences of the University of Medicine and Dentistry of New Jersey School of Public Health. She holds a concurrent appointment at the US Department of Veteran Affairs War-Related Illness and Injury Study Center in East Orange, NJ, where she serves as the risk-communication specialist dealing with deployment-related health risks. Dr. Santos is also the founder and principal of FOCUS GROUP, a consultancy specializing in risk communication, community relations, and health and environmental management. She combines her research and hands-on experience to aid federal, state, and local government agencies and private-sector clients in the design, implementation, and evaluation of health, safety, and environmental risk communication and community involvement programs. Before forming FOCUS GROUP, Dr. Santos served as director of corporate risk assessment services for ABB Environmental, Inc. She also worked for 8 years for Environmental Protection Agency's Region 1 in hazardous-waste management. She conducted research projects exploring how to communicate the results of health studies to community members, including low-literacy audiences, and methods for evaluating stakeholder involvement programs. Dr. Santos has a PhD in law, policy, and society from Northeastern University.

Richard Sharp is director of bioethics research at the Cleveland Clinic. Before joining the Cleveland Clinic in 2007, Dr. Sharp taught bioethics at Baylor College of Medicine and directed the Program in Environmental Health Policy and Ethics at the National Institute of Environmental Health Sciences. His research examines the promotion of informed patient decision-making in clinical research, particularly research that involves genetic analyses. Dr. Sharp is a member of the Ethical, Legal, and Social Implications of Human Genetics Study Section in the Center for Scientific Review of the National Institute of Health. He received his PhD from Michigan State University.

Gina Solomon is the deputy secretary for science and health at the California Environmental Protection Agency (CalEPA). Before joining CalEPA in May 2012, she was a senior scientist at the Natural Resources Defense Council and a clinical professor of medicine at the University of California, San Francisco (UCSF), where she was also the director of the UCSF occupational and environmental medicine residency program and the associate director of the UCSF Pediatric Environmental Health Specialty Unit. Her work has included research on asthma, climate change, and environmental and occupational threats to reproductive health and child development. Dr. Solomon serves on the Environmental Protection Agency Science Advisory Board and on the editorial board of

Environmental Health Perspectives. Dr. Solomon was a member of the National Research Council Committee on Toxicity Testing in the 21st Century. She received her BA from Brown University, her MD from Yale School of Medicine, and her MPH from the Harvard School of Public Health.

Justin G. Teeguarden is a senior scientist in biologic monitoring and modeling at Pacific Northwest National Laboratory. He previously served as chair and president-elect for the Dose–Response Specialty Section of the Society for Risk Analysis. He also served as a member of the Environmental Protection Agency (EPA) Science to Achieve Results (STAR) grant-review panel on computational toxicology. In 2003, Dr. Teeguarden received an award from the Risk Assessment Specialty Section of the Society of Toxicology for the Best Published Manuscript Advancing the Science of Risk Assessment. His current research involves developing an integrated systems-biology–directed research program on effects of particulate matter on respiratory health. He continues to consult both for EPA and for private companies on developing and applying physiologically based pharmacokinetic models and other dosimetry approaches supporting risk assessment. He received his PhD in toxicology from the University of Wisconsin–Madison.

Duncan C. Thomas is the director of the Biostatistics Division of the Department of Preventive Medicine of the University of Southern California and holds the Verna Richter Chair in Cancer Research. Dr. Thomas was codirector of the Southern California Environmental Health Sciences Center (funded by the National Institute of Environmental Health Sciences) and is director of its Study Design and Statistical Methods of Research Core. His research interests include the development of statistical methods in epidemiology, with emphasis on cancer epidemiology, occupational and environmental health, and genetic epidemiology. He is also a senior investigator in the California Children's Health Study. He has published more than 200 peer-reviewed journal articles in those fields and is the author of *Statistical Methods in Environmental Epidemiology* (Oxford University Press, 2009). Dr. Thomas is a Fellow of the American College of Epidemiology and a past president of the International Genetic Epidemiology Society. He has served as a member of National Research Council committees to review radioepidemiology tables, the biologic effects on populations of exposures to low levels of ionizing radiation (BEIR V), and improving the presumptive disability decision-making process for veterans. He was a member of President Clinton's Advisory Committee on Human Radiation Experiments. Dr. Thomas received his PhD in epidemiology and health from McGill University and his MS in mathematics from Stanford University.

Thomas G. Thundat is a Canada Excellence Research Chair professor at the University of Alberta, Edmonton, Canada. Until recently, he was a UT-Battelle/ORNL Corporate Fellow and the leader of the Nanoscale Science and Devices Group at the Oak Ridge National Laboratory (ORNL). He is also a research professor at the University of Tennessee Knoxville; a visiting professor at the University of Burgundy, France; and a Distinguished Professor at the Indian Institute of Technology, Madras. He received his PhD in physics from the State University of New York at Albany in 1987. He is the author of over 263 publications in refereed journals, 45 book chapters, and 32 patents. Dr. Thundat is the recipient of many awards, including the US Department of Energy's Young Scientist Award, R&D 100 Awards, the ASME Pioneer Award, the *Discover Magazine* Award, FLC Awards, the *Scientific American* 50 Award, the Jesse Beams Award, the Nano 50 Award, Battelle Distinguished Inventor, and many ORNL awards for invention, publication, and research and development. Dr. Thundat is an elected Fellow of the American Physical Society, the Electrochemical Society, the American Association for the Advancement of Science, and the American Society of Mechanical Engineers. Dr. Thundat's current research focuses on novel physical, chemical, and biologic detection using micro and nano mechanical sensors. His expertise includes the physics and chemistry of interfaces, biophysics, solid–liquid interfaces, scanning probes, nanoscale phenomena, and quantum confined atoms.

Sacoby M. Wilson is an assistant professor at the Maryland Institute for Applied Environmental Health of the University of Maryland. Dr. Wilson formerly was at the University of Michigan in the Robert Wood Johnson Health and Society Scholars Program, where he developed a research agenda examining built environments, planning, and health disparities. His research interests include the intersection of environmental and social determinants of health and health disparities, the effects of the built environment on vulnerable populations, spatiotemporal mapping of social and environmental phenomena, community-driven environmental-justice research on potential environmental public-health consequences, and geographic information system–based exposure assessment. Dr. Wilson received his PhD and MS in environmental health sciences from the University of North Carolina at Chapel Hill.

Appendix B

Statement of Task

A National Research Council committee will develop a long-range vision for exposure science and a strategy with goals and objectives for implementing the vision over the next 20 years, including a unifying conceptual framework for advancement of exposure science to study and assess human and ecologic contact with chemical, biologic, and physical stressors in their environments. In developing the vision and strategy, the committee will consider exposure-assessment guidelines and practices used by the Environmental Protection Agency and other federal agencies, the use and development of advanced knowledge and analytic tools, and ways of incorporating more complete understanding of exposure into risk assessment, risk management, and other applications for human health and ecologic services. The study will focus on the continuum of sources of stressors, their fate in or changes in the environment, human and ecologic exposure, and resulting doses or other relevant metrics that are relevant to outcomes of concern. The committee's report will potentially be a companion document to previous National Research Council reports such as *Toxicity Testing in the 21st Century: A Vision and a Strategy* and *Science and Decisions: Advancing Risk Assessment*.

Specific issues may include:

- Factors that affect relationships among stressors (chemical, biologic, and physical) and exposed organisms along the continuum from sources to doses in humans, including susceptible individuals or populations, and from sources to ecosystems.
- Innovative approaches for characterizing aggregate and cumulative exposure to various mixtures of stressors via multiple pathways.
- Enhancement of predictive and diagnostic modeling (including probabilistic modeling) in exposure science to reduce uncertainties in risk assessment, risk management, and assessment of mitigation effectiveness.

- Development or improvement of measurement and monitoring methods and interpretive tools (such as informatics) to provide data fundamental to exposure science.
- Exposure metrics (based on prediction and diagnosis) and exposure indicators (based on observations) for assessing the effectiveness of risk management and other decision-making.
- Approaches to increase the usefulness of data from biomonitoring or environmental monitoring in developing risk assessments and related public policies.
- Identification of exposure aspects among humans and other organisms that do not readily lend themselves to inclusion in a unifying conceptual framework for exposure science.
- Key research and development needs for advancing exposure science.
- Educational approaches for training future exposure scientists.
- Communication approaches for conveying exposure-related information to policy-makers and others.

Appendix C

Concepts and Terminology in Exposure Science

Exposure science and allied public-health disciplines have been challenged for more than 2 decades by inconsistent definitions and applications of the terms *exposure* and *dose*. From a strictly observational standpoint, three descriptors characterize the contact between a stressor and a receptor: amount (for example, concentration, mass, and energy), duration (for example, exposure period, duration, and frequency), and location within the system (for example, inhaled air, skin, or target tissue). In addition, some description of how endogenous factors may affect dose is important (for example, inhalation rate is affected by level of physical activity).

The terms external exposure and internal and target-site exposure are elements of the source-to-outcome continuum (see Figure C-1) used in the environmental-health field. The core concept is that the different levels of biologic or environmental organization are separated by barriers to transport or transport processes that need to be accounted for in understanding or describing the relationship between exposure measures on each side of a boundary or level of organization.

From Figure C-1 it is clear that the closer measures of exposure are to the target site for the outcome being examined, the greater is the utility of the data for assessing effects of specific stressors. Conversely, the closer measures of exposure are to environmental concentrations the greater the utility of the information for source emission assessment and control (Figure C-1). The choice of exposure measure is based on the goal of the study or intended use of the data but should always be selected according to what information best minimizes confounders and supports the study's goal or hypothesis. Exposures at any level can be related conceptually and mathematically to exposures at any other level or to dose. On the basis of the above discussion, Figure C-2 (modified from Figure 1-2) provides a framework that more directly characterizes the theoretical and data-collection efforts of the field of exposure science. To accommodate recent advances in biologic monitoring, Figure C-2 contains the term *Internal*

Exposure instead of *Dose* in the "Exposure" box. This modification is intended to address the importance of measuring and quantifying exposure within the organism, but at least a level of organization away from the target site, for example, a specific tissue, or cell, in an organism or a specific compartment of an ecosystem.

The use of the term *internal exposure*[1] is potentially a major shift for the field in that it can be used both in conceptual and theoretical discussions and in experimental design to characterize the processes associated with exposure biology. The quantitative definition of internal exposure is the same as originally discussed by Lioy (1990) and others, but it was described as an internal dose. As the field moves forward, the internal-exposure values can help to establish coherence in the quantitative units that are used to describe the exposure values associated with different routes of entry to the target (for example, mg/kg/day), whether human or ecologic. Therefore *internal exposure* links the internal-marker measurements of exposure (for example, blood and urine) directly to traditional external measures of exposure, and these in turn can be linked to a dose that is described for toxicologic sites of action or for clinical analyses.

FIGURE C-1 Another view of the source-to-outcome continuum for exposure science. Exposure science can be applied at any level of biologic organization: the ecologic level, the community level, or the individual level—and within the individual at the level of external exposure, internal exposure, or target-site.

---

[1]Internal exposure is defined as the contact between an agent or a receptor one level of physical or biologic organization past the external boundary toward the target site.

Appendix C 195

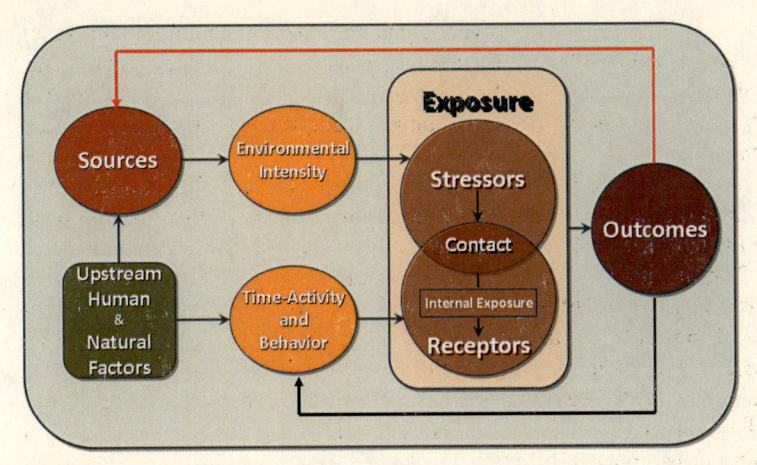

FIGURE C-2 Core elements of exposure science. This figure is modified from Figure 1-2, with *Dose* being replaced with *Internal Exposure*. The term *environmental intensity* is used, because some stressors, such as temperature excesses, cannot be easily measured as concentrations.

REFERENCE

Lioy, P.J. 1990. Assessing total human exposure to contaminants: A multidisciplinary approach. Environ. Sci. Technol. 24(7):938-945.